SATELLITE MONITORING OF
INLAND AND COASTAL WATER QUALITY
Retrospection, Introspection, Future Directions

The cover illustrates the sum of the MSS bands 4, 5, and 6 recorded during an ERTS-1 overpass of the central basin of Lake Erie in March, 1973. These were obtained when the author was establishing an environmental optics/remote sensing program at the National Water Research Institute, Burlington, Ontario. This historical figure illustrates one of the earliest attempts (by an end-user) at digital analyses of satellite data over optically complex inland waters. Seven-track electromagnetic data tapes, a cumbersome mainframe computer, mosaics of printed paper outputs, computer symbolism, quasimanual pixel contouring, Letrafilm, Letratape, pens, colored inks, vellum, and painstaking artwork combined to generate aquatic "maps" from an ERTS "paint-by-numbers" kit. Despite primitive data-handling techniques, the image illustrates the spatial distribution of turbid flow zones; sediment density; north shore–south shore disparities of spring littoral run-off; distinct geophysical gyres and plumes; and gradations in aquatic clarity as water transports from the moderate depth middle basin to the much deeper eastern basin. An early all-encompassing snapshot of lake-wide, dynamics-related phenomena.

SATELLITE MONITORING OF INLAND AND COASTAL WATER QUALITY
Retrospection, Introspection, Future Directions

Robert P. Bukata

National Water Research Institute
Ontario, Canada

CRC Press
Taylor & Francis Group
Boca Raton London New York

CRC Press is an imprint of the
Taylor & Francis Group, an **Informa** business
A TAYLOR & FRANCIS BOOK

First published by Taylor & Francis

Published in 2005 by
CRC Press
Taylor & Francis Group
6000 Broken Sound Parkway NW, Suite 300
Boca Raton, FL 33487-2742

First issued in paperback 2019

No claim to original U.S. Government works

ISBN-13: 978-0-367-45413-5 (pbk)
ISBN-13: 978-0-8493-3356-9 (hbk)

Library of Congress Card Number 2005041711

Library of Congress Cataloging-in-Publication Data

Bukata, R. P.
  Satellite monitoring of inland and coastal water quality : retrospection, intropection, future direction / Robert P. Bukata.
       p. cm.
  Includes bibliographical references and index.
  ISBN 0-8493-3356-3 (alk. paper)
  1. Water quality--Remote sensing. I. Title.

  TD367.B85 2005
  363.739'463--dc22   2005041711

Taylor & Francis Group
is the Academic Division of T&F Informa plc.

Visit the Taylor & Francis Web site at
http://www.taylorandfrancis.com

and the CRC Press Web site at
http://www.crcpress.com

Dedication

To Gail…and the day you came along

Preface

Wakey-wakey!

Frivolous wake-up call, with origin somewhere in antiquity; opening bellow by Billy Cotton on his BBC radio program, 1940s to 1950s

A warning and wake-up call serve as a prelude to *Satellite Monitoring of Inland and Coastal Water Quality: Retrospection, Introspection, Future Directions*, a science monograph written because of a passion for space monitoring of optically complex natural waters and the frustrations that arise from this passion not being shared by the targeted end-usership of the water quality products that emerge from that passion. Although this book was written around a spectrum of thoughts and emotions that pervade (and often torture) the mind of the author, he contends (and hopes) that these thoughts and emotions are not merely self-imaging, but rather that they fairly and genuinely portray the passions and frustrations of colleagues and friends with whom he has shared all or parts of the greater than three-decade odyssey comprising the remote sensing of optically complex inland waters. Therefore, throughout this book, the author will, although somewhat presumptuously, write in the conversational first-person plural "we" (as well as in the detached and impersonal third person) and forsake the egoistical first-person, singular "I."

Retrospection and introspection — as well as personal (and thus controversial) conjurations arising therefrom — will form the basis of our attempts to probe reasons for the uneasiness that appears to shroud the acceptance of remote sensing of inland and coastal water quality as a genuine, respected, and valued component of environmental monitoring. Inarguably, introspection of an impassioned object and retrospection of probable causes for the uneasiness in which it is envisaged could result in indiscriminate "finger-pointing." Although fingers will be pointed, this is not intended to be indiscriminate or to cast aspersions. Science and technology are independent and pure entities that search for truth and inventiveness. However, they must adjust to and coexist with the public that supports them. As such, scientific truth and technological invention become modulated by the mind-sets (and thus the whims and caprices) of influential special-interest groups within that public.

Elected public representatives, along with the ensuing government policies and funding priorities, also become modulated by these special-interest mind-sets. Not unexpectedly, such mind-sets are often at variance with the independent search for truth and invention aspects of science and technology. Environmental monitoring from space was born during a disruptive period of cultural, social, and political unrest. Attempts at establishing "all-inclusive" solutions to a variety of issues that were disturbing to a variance of public mind-sets led to unfortunate combinations of well-intentioned, as well as self-serving, mistakes. The resulting discomfort that encumbered acceptance of remotely monitoring the quality of inland waters might well have been an unavoidable consequence of those mistakes.

Discussions of mistakes invariably result in real and perceived accusations of consequence, if not outright blame. Undoubtedly, interpretations and opinions presented herein will be contrary to those of some readers. Although controversy is anticipated, no calumny is intended. The principal goal is the exit from the discomfort plaguing the space monitoring of inland and coastal waters. Any real or imagined problem generally becomes more solvable once probable root causes can be defined. The root causes of societal reluctance to accept products resulting from scientific research must invariably be ferreted from the transitory secular dynamics that have shaped and continue to shape the history of science and society. Perhaps, more fittingly, they should be ferreted from conflicts between science and populist-culture perceptions of science.

From its inception, Darwinism has been highly controversial and has lacked societal acceptance. It took over 200 years for gravity to be accepted as scientific law. Einstein's relativity is still regarded as a theory. The Flat Earth Society has never lacked for new membership. Cold fusion was readily accepted (mercifully, for only a short period of time) as a viable source of energy. Within a scant number of years, anthropogenic carbon dioxide gas emission as a motor force for climate change is becoming accepted as fact, gospel, and doctrine; despite its convolutions of myth and logic, this idea garners large-scale public and media support. And so on and so on.

Perhaps, no greater burden exists than that of an ascribed great potential. Such was the burden placed upon remote sensing of Earth's natural resources from space that became a reality ~30 years ago. Well-intentioned, although irresponsible, promotion by overly zealous proponents of an as yet untried venture resulted in volumes of inappropriate literature that fostered understandably warranted skepticism in potential users of this novel approach. The scientific community itself displayed continual reluctance to accept remote sensing as a worthwhile activity. This gave rise to an unfortunate pseudoscientific entity with the premise that validity could be ascribed to remote sensing by willy-nilly correlations of remotely acquired data with virtually any and every environmental variable measured on the ground. Such actions imposed acceptance criteria on remote sensing that relied upon a suspension of disbelief. Proliferation of the self-deprecating term "ground truth" (incredibly, still within the active vocabulary of the remote sensing community) served only to suspend the suspension of disbelief.

Subsequent sound science by perseverant dedicated workers managed to keep remote sensing from being ultimately "sent to Coventry" by such inappropriate salesmanship. However, despite such sound science, remnants of this initially deserved suspect reputation persist. Acceptance of remote sensing is further handicapped by the routinely prohibitive costs associated with execution of remote sensing missions, as well as with obtaining the three decades of archives of unanalyzed and financially inaccessible time-series satellite data.

Remote sensing is dramatically underutilized in ecosystem monitoring and assessment. Operational remote sensing of inland and coastal water quality is virtually nonexistent. Although past decades have seen growing interest and comforting advances in acceptance of remotely sensed oceanic and terrestrial products, the future of remote sensing of inland and coastal waters continues to be jeopardized by steadfast end-user disinterest in its products. Advances in sensors appropriate to monitoring the quality of optically complex waters remotely continue. Satellites capable of water quality monitoring have recently been launched or are planned for near-future launch. Thus, despite decades of technology push, the remote sensing of inland and coastal waters is still waiting for end-user pull. Does the ascribed great potential still exist?

If we make the not unreasonable assumption that an ascribed great potential does not necessarily preclude a great (or at least adequate) potential, then such ascribed potential may be considered as merely an as yet unfulfilled potential. Adopting this assumption, the remote sensing of inland and coastal water quality will be accepted here as an underutilized valuable component of environmental monitoring, the value of which has yet to be compellingly elucidated to influential end-users. We suggest that this compelling elucidation become the responsibility of those whose talents and dedication have resulted (and are continuing to result) in the generation (and on-going upgrading) of these water quality products. A popular dreamscape has emerged from the philosophical adage that players will be attracted to a field upon which they can practice their craft. We have built such a field — thus far, we are the only ones playing on it!

Remote sensing science and remote sensing technology will probably never be considered optimal by the talented teams of scientists and technologists that are expanding them. Nor should they be so considered. The insatiable quests for the ultimate model, the ultimate algorithm, and the ultimate sensor are invaluable motivators. Although, within such insatiable quests, remote sensing technology will, understandably, always lead remote sensing science, the existing remote sensing science (admittedly less than optimal) is nonetheless extremely good.

Despite occasional pratfalls, remote sensing scientists have made creditable and enviable advances in explaining and mathematically formulating the physics and biophysics of the radiative transfer processes occurring within aquatic media. These have been made initially for mid-oceanic waters and subsequently for the more optically complex waters adjacent to or encapsulated by dominant land masses. That sound and defendable science is essential to remotely sensing water quality is an irrefutable given. That practical and defendable applications of the products resulting from such science are essential for the future of remotely sensing water quality is also an irrefutable given.

In a Utopian universe, these irrefutable givens would be in perpetual harmony. In the non-Utopian universe that we inhabit, however, these irrefutable givens are often in conflict. Conflicts may be resolved by only three possible means: warfare, capitulation, or mutually acceptable compromise. In this book, we endorse the latter and, consequently, advocate a compromise between defendable science and defendable usage of that science. Hopefully, the demands placed upon remotely sensing inland and coastal water quality and the applications that will be presented to potential influential end-users of that remote sensing will respect and adhere to that compromise.

Although we advocate a defendable compromise between science and its usage, we irrevocably stress that the credibility of the science must remain untarnished, i.e., the scientific integrity must not be compromised. Therefore, if the conflict between defendable science and defendable usage of science cannot be resolved by compromise, capitulation by the usage must logically follow. If only the early prophets of the "great potential" had seen this!

Of necessity, this book addresses how science, while maintaining its mandatory mantle of truth and fidelity, must coexist and cope with the imperfections and demands of the society in which it is entrenched. Thus, this book is not intended to be a major treatise on aquatic optics and the remote sensing of inland and coastal waters *per se* (although Chapter Two and Chapter Three present an anecdotal history followed by a detailed discussion of the philosophy and bio–geo-optical modeling that made satellite monitoring of the quality of optically complex waters an operational reality). Excellent historical and recent textbooks and monographs have been written detailing the radiative transfer processes that dictate the energy transfer within natural water bodies as well as the energy transfer across the air–water interface. Readers should capitalize upon these books' availability. They should also take advantage of historical and recent publications written by the authors of such textbooks and monographs, as well as by the workers whose cited research is responsible for the intelligence contained therein.

However, although this book is not intended to be a major treatise on aquatic optics, it *is* intended that this book present, in textbook format, the science necessary to understand how water color can be converted into water quality. It is also intended that this book serve as a guide to illustrating how such conversion can produce products of consequence to environmental monitoring of optically complex inland and coastal waters. Sadly, the excellent and innovative scientific progress made in the optics and bio-optics of these so-called case 2 waters, although appreciated and advanced within the scientific remote sensing community, is neither appreciated nor recognized much beyond that scientific community.

In general, remote sensing (terrestrial, aquatic, and wetland) is an underutilized tool within the environmental monitoring tool kit. Remote sensing of inland and coastal waters is a dramatically underutilized tool that may, in fact, not even reside within the environmental monitoring tool kit. The plight of remotely sensing inland and coastal waters is, quite simply, the absence of an influential end-usership for its water quality products. Without such an end-usership, the future of remotely monitoring optically complex waters becomes bleak. In this book, we attempt to take a realistic (although admittedly sympathy-laden) look at this end-user disinterest in water quality products. Why does it exist? How may it be abated? What possible synergies need to be activated among technologists, scientists, private sector entrepreneurs, water quality managers, and political policy-makers?

The drama that is acceptance of remote sensing of water quality as a valuable component of environmental monitoring protocols is being played upon a stage that faces a disinterested yet surprisingly critical audience. This audience is not fully aware of the drama or even that it is part of the cast. In order to produce this drama sensibly, we have elected to target environmental decision-makers and political policy-makers as the audience for the book. This audience must be made aware that they are, indeed, a vital part of the cast, and that the drama is worth producing.

Perhaps even more importantly, however, this "guidebook" is directed to veteran and rookie scientific users fully aware of the importance of remote determinations of water quality, who have shared in the frustrations regarding lack of nonscientific end-user interest. Very important members of this scientific group are the undergraduate and graduate students. We request that they and their academic mentors focus on environmental applications of aquatic products resulting from prior and on-going research in inland and coastal aquatic optics.

Communication, obviously, is a major hill that must be climbed. Workshops, symposia, conferences, planning sessions, and proposals have formed the history of the past almost four decades of remote sensing promotion. These have all been attended and/or generated by various combinations of providers, modelers, vendors, users, and potential users of remotely sensed data. Included within this history is a continuum of pilot studies designed to illustrate how remote sensing can provide cures to a variety of real, perceived, and anticipated environmental ills. After all of this, "the great potential" remains unfulfilled. Only four conclusions emerge from the lack of realization of potential and allegedly fraudulent promises:

- The promises were unrealistic.
- The remote sensing was unrealistic.
- The assembled gatherings were inappropriate.
- The dialogue was unrealistic.

Arguments can be presented to support each conclusion. However, because the first three conclusions may arise from the last one, supplying realistic dialogue appears to be the best hope for the great egress from the plight of remote sensing of inland and coastal aquatic resources.

Specific reasons for over three decades of unfulfilled promises and potential are numerous and varied. Some are obvious, some are speculative, some are real, and some are perceived; all are polemic. The net result, however, has been a disparity between the promise of remote sensing and the actual delivery of what influential end-users wanted in the areas of environmental decision-making and political policy-making. It would appear fairly straightforward simply to ask these end-users what they want and then have the remote sensing community supply it. That it has not been straightforward is the impetus of this book. Existing problems (some scientific, some social, and some financial — all perverse) conspire to prevent successful scientific conversion of remotely sensed data into inland and coastal water quality information from forming a valid component of environmental monitoring networks. Possible solutions to these problems through obligatory compromises between defendable science and defendable usages of science form an underlying theme of this book.

As a guide to possible compromises that would provide environmental decision-makers and policy-makers with inland and coastal water quality information, we have used the research and science agenda of Environment Canada as a template for generic environmental interests and concerns. Using these priorities as a guide and drawing from an as yet limited sample case, examples of airborne and satellite water quality products will be presented. Hopefully, these will satisfy the compromise criteria and provide a compelling case for the incorporation of aquatic remote sensing into networks of environmental monitoring.

Throughout time, it has been acknowledged that "a picture is worth a thousand words." Assuming such an adage is inarguable (i.e., a picture and 1000 words are indeed interchangeable), it imposes an obligation on both the picture and its viewer. First, the picture must exist and warrant 1000 meaningful words. Second, the viewer must possess an active vocabulary of 1000 meaningful words. The active vocabulary of the aquatic bio-optical scientist that produces the remote sensing picture is not at issue. The vocabulary of an aquatic bio-optical scientist viewing the picture is also not at issue because the producer as well as the scientific viewer is working from the same dictionary.

However, the active vocabulary (but different dictionary) of the intended environmental policy-making end-user viewing these scientifically descriptive pictures does become at issue. Satellite images and satellite data products generated from selected space agency algorithms are becoming readily available on the Internet; however, a large pool of compellingly illustrative (to nonscientifically trained viewers) images from which to choose does not, as yet, exist. This limited availability of "pictures" has therefore often forced the "thousand words" option upon a sizeable fraction of the applications and illustrative examples discussed in Chapter Four and Chapter Five. The lacuna of compelling language-translatable "pictures" is an issue that must be overcome and is another impetus of this book. As illustration of this lacuna, consider the artist who has painted a picture of a jabberwocky and begins his or her 1000 meaningful words with:

'Twas brillig, and the slithy toves/Did gyre and gimble in the wabe/All mimsy were the borogoves/And the mome raths out-grabe.

From *Through the Looking Glass and What Alice Found There*, **Lewis Carroll, 1872**

Although our selected illustrations are representative of waters on a global scale, we will draw from early and recently initiated local programs and activities that we feel afford a luxury of optimism that the future might be a little less frustrating than the past. In the closing chapters of this monograph, we will discuss developing and on-going work within

- Environment Canada's National Water Research Institute (NWRI)
- Tripartite NWRI, Institute of Ocean Sciences (IOS), and G.A. Borstad Associates Ltd. projects under the aegis of the Canadian Space Agency (CSA)'s Earth Observation Applications Development Program (EOADP) and Government Related Initiatives Program (GRIP)
- Ocean color collaborations of Bedford Institute of Oceanography (BIO) and Dalhousie University, Nova Scotia
- Water research programs of Commonwealth Scientific and Industrial Research Organization, Australia (CSIRO)
- The consortium of academic and federal agencies and industries in Minnesota, Wisconsin, and Michigan under the initial rubric of NASA's Regional Earth and Applications Center (RESAC) and now extended under NASA's Affiliated Research Center (ARC)
- The Naval Research Laboratory, Stennis Space Center (NRL/SSC), in Mississippi
- The NOAA CoastWatch Program at the Great Lakes Environmental Research Laboratory (GLERL), Michigan

Writing a textbook/guidebook based upon a collage of science, applications of that science, and public acceptance of those applications, as well as retrospection, introspection, and possible future directions of such a collage, can be a difficult undertaking. Directing such a text to a collective audience of diverse backgrounds, interests, and responsibilities in a manner that addresses their common ground presents further difficulty. Mathematical formulations of radioactive transfer theory can become confounding to end-users who have no background in spectro-optical research and modeling.

In an attempt to make this text less confounding to a varied, although related, audience, we have chosen to provide anecdotal as well as scientific details to review how aquatic optics has made space monitoring of optically complex water quality possible and how, despite problems that still require attention, recent science advancements have added authority and value to water quality monitoring. Historically, the relationship of science with society has adhered to the ever changing dynamics of science and populist culture, science and policy, and science and applications. These dynamics have had major impact on the evolution of environmental monitoring from space. Therefore, we have attempted to write this book as a reader-friendly dialogue dealing with events and prevalent attitudes so as to show their interaction.

If all the problems obstructing acceptance of remote determinations of water quality as a valid environmental monitoring practice were insignificant, solved, or, better yet, nonexistent and there were influential end-users of remotely sensed inland and coastal water quality products, there would be no need for this book. Hopefully, there will not be a need for another. In its place, a chronicle of success stories would, however, be most welcome. This chronicle is the charge directed to the intended readership.

Robert P. Bukata

To you from failing hands we throw/The torch; be yours to hold it high.

From *In Flanders Fields*, John McCrae, 1915

Acknowledgments

To be, or not to be: that is the question:/…suffer/The slings and arrows of outrageous fortune,/…bear the whips and scorns of time,/…makes us rather bear those ills we have/Than fly to others that we know not of

From *Hamlet, Prince of Denmark*, **Act 3, Scene 1,**
William Shakespeare, 1603

Allegations of paranoia notwithstanding, being a remote sensing enthusiast and, furthermore, admitting that you are one have not always been easy. Unlike physics, chemistry, biology, sociology, medicine, music, literature, and other sciences and arts, remote sensing is not a discipline; it is an activity. Compounding the problems of its nondisciplinary status, remotely sensing the Earth's biome becomes, of necessity, a *multidisciplinary activity.* By contrast, remote sensing of the interplanetary medium (i.e., exclusive of planetary life forms) is essentially a single-disciplinary (physics) activity and thus qualifies as a discipline (space science).

Environmental remote sensing was an out-of-wedlock child of space science parents. Legitimatizing and raising this love-child has been the goal of its space-science parents, as well as of disciplinary foster parents who have added their interests and talents to the cause over the years. Remotely monitoring mid-ocean water quality (thereby legitimatizing *that* activity) advanced fairly rapidly due to (1) the global teams of talented researchers in oceanographic institutes performing definitive ocean optics research for decades prior to satellite launchings and (2) ongoing research in ocean–atmosphere interaction being a recognized scientific priority area for explaining the cyclic nature of the Earth's climate.

Understandably, therefore, oceanography readily qualified as a fit foster parent. However, limnology, whose cumulative domain is essentially negligible when compared to the cumulative domain of oceanography, did not. Thus, the handful of inland water remote sensing scientists applying for foster parent status who were, understandably, critical (as well as envious) of the comparatively easy-come successes of oceanographic colleagues were often dismissed as being troublesome, petulant, or even heretical by many within the remote sensing community. Thus, a three-decade odyssey (such an extended time-frame was not foreseen) was begun, first to show that limnology was indeed a fit foster parent, and then to show that space monitoring of inland water quality was of consequence to science and to environmental stewardship.

As a space physicist and an alleged "heretic" concerned with remote sensing of inland water quality right from the launch of ERTS-1, I am grateful that I did not need to make the three-decade odyssey alone. From the outset, two colleagues and friends took initial tentative baby steps with me: John H. Jerome and J. Edward Bruton quickly became instrumental in turning those steps into strides. John remains vital to the odyssey; through a cruel inescapable destiny, Ed was suddenly and tragically taken from it. I am grateful to many people who have had an impact on the odyssey — directly through personal contact, collaboration, and support or indirectly through my respect for their exceptional work.

In addition to John and Ed, a woefully incomplete list of colleagues who must be acknowledged for my not having chosen the "or not to be" option to Hamlet's question and thus walking away from aquatic remote sensing and returning to the sanctuary of the established field of physics includes Richard A. Vollenweider; Clifford H. Mortimer; William S. Haras; John T.O. Kirk; Jeff Whiting; Shubha Sathyendranath; Janet W. Campbell; Jim F.R. Gower; Gary A. Borstad; Robert C. Wrigley; Allan Hollinger; John R. Miller; Robert A. Arnone; Kirill Ya. Kondratyev; Dimitry V. Pozdnyakov; George A. Leshkevich; Barry M. Lesht; William G. Booty; Robert A. Stavn; Jean-Michel Jaquet; and senior management of the Canadian Space Agency. I trust that my decision to use the first person plural "we" as opposed to the first person singular "I" in this book will meet with their approval, if not their total agreement.

I thank Tony Moore, senior editor at Taylor & Francis, for the invitation to write this book. I also thank Arnold G. Dekker and Thomas M. Lillesand for discussions and encouragement during its initial preparation. Their programs at CSIRO and RESAC/ARC, respectively, bode well for water quality monitoring. Very special thanks are extended to Arnold for his and my being able to establish an open forum for scientists, space agency managers, providers of aquatic products, and potential users of those products at the 2001 Alliance for Marine Remote Sensing (AMRS) inland and coastal waters workshop in Wolfville, Nova Scotia. In some ways, this book is an outcrop of that forum. I am indebted to my colleagues for allowing me to use illustrations from their authoritative work in this conglomerate of retrospection, introspection, and prophecy.

On a highly personal note, the dearest companions on any odyssey are the full-time, sympathetic, supportive, and understanding loved ones who have willingly signed on board for the duration. I was/am blessed with an enviable passenger list: my comfort zone, bride, and touchstone, Gail, and our thoughtful, perceptive daughter, Sheri, and son, Andrew, who are now on odysseys of their own blessed with equally enviable passenger lists.

Thank you. Your presence and actions continuously demonstrated that the truly important aspects of the journey had absolutely nothing whatsoever to do with inland water quality monitoring.

R.P.B.

About the Author

Robert Peter Bukata received his doctorate in physics and mathematical physics in 1964 from the University of Manitoba where he devised a pair of bidirectional, mutually perpendicular, high-energy particle telescopes rotating in opposition to the Earth's rotation to discover a source of very high-energy galactic radiation in the constellation Aquila. He spent 7 years on the faculty of the Southwest Center for Advanced Studies in Dallas, Texas, as coprincipal investigator of cosmic ray studies aboard the NASA deep-space probe missions Pioneers 6 through 10 and the Earth-orbiting satellites Explorers 35 and 41. His Dallas team's Pioneer cosmic ray sensors (monitored by NASA throughout its lunar missions to ensure that Apollo launch schedules would not coincide with space radiation hazards to astronauts) were the first to observe directly the abundance of solar flares generated during the least active segment of the 11-year solar cycle. He and his colleagues utilized *in situ* measurements of the anisotropies in direction arrivals of solar proton and alpha particles at the suite of solar-orbiting Pioneer probes to conceptualize a filamentary microstructure model for the interplanetary magnetic field and to provide fundamental information on electrodynamic processes (solar flares; "behind the sun" solar activity; M-region magnetic storms; energetic storm particle events; corotating Forbush decreases; modulation of galactic radiation) that govern propagation of solar particles within the inner solar cavity. He also utilized high-altitude balloons and the world-wide ground-based neutron monitor network to explain production and behavior of secondary cosmic radiation within the Earth's atmosphere.

In 1971, Dr. Bukata designed a program of airborne remote sensing of cultivated crop bioproductivity for the Provincial Government of Manitoba. In 1972, he joined Environment Canada's National Water Research Institute (NWRI) in Burlington, Ontario. Here, he established a novel program of limnological spectro-optics research resulting in bio–geo-optical models to determine inherent optical properties (optical cross section spectra) of organic and inorganic color-producing agents comprising optically complex inland waters, and to extract the concentrations of these coextant aquatic agents from remote full-spectrum measurements of water color. At NWRI, Dr. Bukata has served in a variety of management capacities, including head of remote sensing, head of environmental spectro-optics, chief of physics, chief of applied physics and systems, and chief of atmospheric change impacts.

Dr. Bukata has held adjunct professorships at several universities in the United States and Canada and has authored over 140 scientific publications and book segments in cosmic ray physics, aquatic optics, remote sensing, limnology, hydrology, UV-radiation, and climate change. He has been scientifically active in Canadian bilateral agreements with Russia and Australia and has received awards and citations for his work in space science and remote sensing. Dr. Bukata is currently working with the Canadian Space Agency, NOAA, and colleagues within Fisheries and Oceans Canada, the private sector, and academia in developing and coordinating space monitoring programs consistent with the mandate of Environment Canada. He is principal author of the CRC Press book *Optical Properties and Remote Sensing of Inland and Coastal Waters.*

Contents

Disclaimer

Introspectives and retrospectives are heavily weighted to hind-casting, thereby resulting in the requirement to live in (or rather relive in) the past. Retrospectives stem from individual recollections, and individual recollections stem from a time series of experiences that are consequences of existing populist culture. Pop culture is forged from interactive admixtures of politics, religion, entertainment, fashion, news media, education, law, sociology, and the ability of these entities to manipulate mind-sets. Music is a powerful force of pop culture and its lyrics provide reasonably accurate, albeit provocative, insight into selective directions of such mind-set manipulation. It was the intention to quote these lyrics throughout the narrative of this monograph to illustrate the ambient environment into which environmental monitoring from satellites was introduced and the role that pop culture (specifically, its perception of science and the role of research) possibly played in the steadfast disinterest (by environmental stewards and policy-makers) in space-acquired inland water quality products. Lyrics of every individual song are individually copyrighted by individual copyright holders. Tracking down these individuals and obtaining their permissions to quote from those lyrics is a formidable and time-consuming task. Consequently, although specific song titles are mentioned, their lyrics, which would have furthered the tone and mood of discussions, are not quoted. The short quoted passages from existing literature or an individual's statement are credited to the source and the author or speaker. The dialogue spoken by King Criswell in *Plan 9 from Outer Space* is its most famous public utterance; however, this was not its first utterance. It was spoken earlier in a promotional movie at the General Motors Futurama exhibit at the 1939 New York World's Fair and, prior to that, at trade union rallies in the 1930s. *Superman* and all related elements are the property of DC Comics.

R.P.B.

chapter one

Navel gazing at remote sensing of inland and coastal waters from space

"The time has come," the Walrus said/"To talk of many things:/ Of shoes and ships and sealing wax —/Of cabbages and kings —/And why the sea is boiling hot —/And whether pigs have wings."

From *Through the Looking-Glass and What Alice Found There*, **Lewis Carroll, 1872**

Realistic self-appraisal is most often agonizingly elusive, due, in no small measure, to the understandable conflict between passion and reason. Nonetheless, for the future of remotely sensing the quality of inland and coastal waters to be evaluated fairly, such self-appraisal becomes mandatory. Appraisals require criteria and criteria imply boundaries and limits. Remote sensing of inland and coastal waters has evolved as synergistic interplay among theoretical and applied modeling, sensor development, and calibration and validation based upon the principles of radiative transfer theory and aquatic optics. Its current status, therefore, is the consequence of synergistic interplay among diverse talents, disciplines, and egos. Because the boundaries and limits have been set by these talents, disciplines, and egos, the desired objectivity of criteria has often given way to subjectivity. Therefore, the following introspection and retrospection while striving for reason, does acknowledge an inescapable degree of passion.

1.1 Environmental monitoring from space

Generally considered to be the observation of a selected target by means of a sensor not in direct contact with that target, remote sensing is a concept that was revisited during the "future shock" vision of environmental deterioration prevalent in the mid 1960s. The era of environmental monitoring from space vehicles may be reckoned to have begun in August, 1972, with the launch of the Earth Resources Technology Satellite (ERTS-1), the first of a series of NASA-planned Earth-orbiting, ecologically oriented satellites. Primary objective of the ERTS-1 mission was to obtain multispectral images periodically over environmental targets and experimentally apply these images to investigations dealing with a variety of variables pertinent to Earth's resources. Thus, scientific and other end-user communities of such remotely acquired environmental data were very early recognized as essential to the continuous execution of this primary objective.

Shortly after launch, however, ERTS-1 was renamed Landsat-1 and the series of Landsat vehicles (Landsat-7 is most current) has provided three decades of time series of some combinations of visible, near infrared, and thermal infrared "snapshots" of environmental change, although within limited spectral resolution. As evidenced by its abrupt name change, however, ERTS-1 was deemed more appropriate for terrestrial than aquatic monitoring. This was due, in part, to:

- The (then) sharply focused interest on agricultural crop diversity and yield, forest clear-cutting, vigor of natural verdure, and terrain identification
- The minuscule reflected visible radiation return from water as compared to the return from land
- Insufficient spectral resolution to define water color
- Insufficient knowledge of the optical properties of natural waters

The 30+ years subsequent to the launch of ERTS-1 have seen proliferation of aircraft and spacecraft sensor systems with ever increasing sophistication in terms of available spectral regions of the electromagnetic spectrum, spatial and spectral sensitivities, and viewing capabilities. Aircraft sensors directed towards terrestrial and aquatic interests have included such imaging spectrometers as the compact airborne spectrographic imager (CASI); airborne imaging spectrometer (AISA); airborne visible/infrared imaging spectrometer (AVIRIS); airborne hyperspectral mapper (HYMAP); and reflective optics system imaging spectrometer (ROSIS). Spacecrafts and sensors exclusively dedicated to the study of ocean color included Seasat; coastal zone color scanner (CZCS); ocean color and temperature scanner (OCTS); sea-viewing wide field-of-view sensor (SeaWiFS); polarization and directionality of Earth's reflectances (POLDER); moderate resolution imaging spectroradiometer (MODIS); medium resolution imaging spectroradiometer

(MERIS); and global imager (GLI). MERIS, however, is the only multispectral satellite sensor that focuses on coastal and inland waters.

The launch of the Earth Observation Platform (EOP-1) with its hyperspectral sensor, Hyperion, ushered in an era of super- and hyperspectral remote sensing that includes scheduled launches of the Australian Resource Information and Environmental Satellite (ARIES); Naval Earth Map Observer (NEMO); Land Surface Processes and Interaction Mission (LSPIM); and Spectral Imaging Mission for Science and Application (SIMSA), among others. However, NEMO is the only hyperspectral satellite sensor specifically configured for coastal and inland water imaging.

Details on these and numerous other satellites and sensors may be found in IOCCG (1998) and on the World Wide Web sites of NASA and other national space agencies. Hilton (1984), Dekker et al. (1995; 2002a), and others have presented reviews of imaging spectrometry theory and its applications to inland and coastal waters.

Clearly, therefore, environmental remote sensing is not restricted by a lack of suitable planned or available satellites and/or sensors. Very briefly, sensors mounted on remote platforms fall into two broad categories: *passive* and *active*. A passive remote sensing device merely responds to the existing environmental condition of the target that it is monitoring. It does not influence the behavior of the target. An active remote sensing device, however, transmits a signal to the target being monitored, stimulates an optical transition within that target, and then measures and records the returned signal. Thus, active devices do influence the behavior of the target and record spectral signatures that would not be recorded if the target were not in an excited optical state. Active remote sensing devices include lasers and synthetic aperture radar (SAR). Passive remote sensing devices include imaging cameras and multispectral radiometers (including visible and thermal). Because this book considers water color, its primary focus will be on passive systems.

Features of the retrospection/introspection on remotely monitoring the quality of optically complex waters presented in this chapter are consistent with positions and views adopted by Bukata et al. (2001b) and Dekker et al. (2001a). Scientific terms encountered in the introspection to follow will be recounted and defined in Chapter Two, which is a brief anecdotal history of the scientific philosophy that led to remote sensing of the Earth from space, and Chapter Three, which is a brief "walk-through" of the science behind the generation of the water quality products. As a further aid and quick reference, a glossary of scientific terms (Appendix B) is located near the back of this book.

1.2   The remote sensing definition of water quality

Many of the Earth's interactive ecosystems have been and are responding to environmental stressors such as

- Changes in land use
- Injections of atmospheric carbon dioxide as industrial by-products
- Deforestation
- Toxic contaminants from farming, manufacturing, and mining practices
- Fluctuations in the cyclic behavior of stratospheric ozone concentrations
- Increased human population in urban centers
- Local disruptions to nitrogen fixation and carbon cycle dynamics
- Year-to-year variations in climate patterns

Understandably, the impacts of such stressors will increase as escalating human population and economic pressures are placed upon the Earth's natural resources. Such cumulative and synergistic impacts may result in serious disruptions to terrestrial and aquatic systems on local to global scales, although their health, life-style, economic, and political consequences must still be considered as speculative and debatable. For example, research into *global warming* (which has become an inappropriate popular synonym for climate change) has suffered from an admixture of scientific facts, nonscientific hypotheses, scientific and societal judgments, preconceived mind-sets from a variety of nonscientifically motivated advocacy groups, and invective emerging from conflicting economic, political, environmental, and media ideologies.

Direct evidence of climate change has been hampered by a small number of facts and a large number of judgments predicated upon very limited information. There is no doubt that natural climate change is occurring (the Earth has gone through many an ice-age to ice-age cycle). There is also no doubt that the Earth is subject to environmental pollution. Unfortunately, an apparently inescapable tendency of well-meaning and self-serving environmental activists alike is the inadvertent or willful confusion of local pollution of local environments with anthropogenic "warming" (or possibly *global cooling*, which was an equally inappropriate popular synonym for climate change in the 1960s and early 1970s) of the global environment. Interestingly enough, water vapor comprises ~97% of the Earth's "greenhouse gases"; carbon dioxide (which is not a global atmosphere toxic pollutant, but rather a necessary nutrient for plant life) constitutes less than 0.05% of the atmosphere, of which only ~3% is from human activity. (This ultimately translates into human releases from natural body functions and industrial pollution of carbon dioxide into the atmosphere on a global basis being responsible for ~0.11% of the atmospheric greenhouse effect.)

Nonetheless, controversial treaties such as the Kyoto Protocol take direct aim at carbon dioxide emissions as local pollution that results in enhanced greenhouse gas control of the global climate (initially as an agent for cooling, currently as an agent for warming). If environmental monitoring from satellites is to contribute realistically to the conflicting admixture of selective science and selective emotion, it should do so in a manner consistent with the 1960s principal objective of the ERTS-1 visionaries. This requires that the past three decades of data from a galaxy of national space agency and private sector environmental satellites (that comprise dispassionate, impressive, although hitherto largely unanalyzed, archives of data) be brought to the foreground as time series utilized in a scientifically sound manner.

Analyses of these data as stand-alone entities may not successfully distinguish between a well-defined natural and a debatably possible anthropogenic change within a local climate. However, such an inability should not become a limiting criterion for satellite monitoring of local environmental change (e.g., water quality) or assessing the impacts of such local change in terms of local environmental stressors (e.g., pollution, land use, water use). We shall re-emphasize this cautionary principle in Chapter Seven.

The wealth of satellite data at thermal, microwave, and radar frequencies notwithstanding, by far the greatest archives of remotely sensed environmental data consist of reflected visible radiation (the so-called "optical data"). Therefore, time series of environmental color become essential for assessing ecosystem change. Radiometric color of a natural water body is governed by the spectral dependencies of in-water absorption and scattering of downwelling global radiation (direct solar and diffuse skylight) by the concentrations of indigenous organic and inorganic color-producing agents (CPAs) coexisting within the water column at the time of the remote observation. Although the principal CPAs of mid-oceanic waters (the so-called "case 1") are phytoplanktonic biota and their co-varying detrital matter, the principal CPAs of inland and coastal waters (mainly representative of the so-called "case 2") include not only phytoplankton and detritus, but also land-derived suspended particulate inorganic matter, colored dissolved organic matter, and (if the water body is shallow and clear enough to allow global radiation to penetrate to the bottom and re-emerge from the air–water interface) the bottom substrate and whatever biological cover it might sustain. These CPAs very rarely display co-variance.

Therefore, in order to utilize water *color* as a measure of water *quality*, an appropriate remote sensing definition of water quality must be presented in terms of these in-water organic and inorganic CPAs. A minimum definition of remotely sensed water quality, therefore, would be the simultaneous inference of the coextant concentrations of chlorophyll *a*, *chl* (quantifiable surrogate for phytoplankton); suspended minerals, *sm* (quantifiable surrogate for inorganic particulates); and dissolved organic carbon, *DOC* (quantifiable surrogate for colored dissolved organic matter). Each of these surrogate CPAs does not represent a single species of phytoplankton, inorganic particulates, or colored dissolved organics, but rather a collection of numerous species that, although representative of their genre, display distinctive spectro-optical interactions with downwelling global radiation. For some shallow eutrophic inland waters, it might also be necessary to define a quantifiable surrogate for suspended organic matter (SOM) as a coextant

CPA. Unambiguously identifying the presence and concentrations of case 2 water CPAs could therefore become a quite formidable task.

1.3   The plight of remotely monitoring case 2 water quality

Years of dedicated physical and bio-optical modeling coupled with *in situ* optical monitoring, sensor development, and laboratory analyses of collected case 1 water samples have resulted in the generation of remotely acquired water quality products that are now routinely used with a reliability generally deemed acceptable by ocean scientists and resource managers. Despite comparable physical and bio-optical modeling and *in situ* monitoring focused on inland and coastal waters, however, end-user disinterest in remotely acquired case 2 water quality products is common and operational applications virtually nonexistent. Possible reasons for such disparity in acceptance include

- The perception that, although remote sensing might be essential to conventionally unattainable ocean waters, it is not essential to conventionally attainable inland and coastal waters (at least those in proximity to highly populated areas)
- The recognized major roles of oceans in global climatology
- The optical simplicity of case 1 waters ensuring an accuracy (in absolute, but not necessarily relative values) for ocean-derived remote sensing products that would exceed the accuracy of such products derived for optically complex case 2 waters

Because the second point is factual and the third is quasifactual, it would appear that the most profitable approach to attaining increased acceptance of case 2 water quality products would be through the first point — namely, illustrating to the end-user community the need to monitor inland and coastal water quality remotely. This, essentially, will be the main theme and purpose of this book. Quite simply, the plight of remotely monitoring case 2 water quality is its lack of acceptance by influential end-users outside the scientific community producing them. The dialogue and illustrations to follow in this book will attempt to explain how this plight arose, why it still exists, and how it might possibly be assuaged.

It is important to realize the nature of the water quality products that can emerge from remote measurements of case 2 water color and the information that may be garnered from such products. It is equally important to recognize precisely the users of remotely acquired data and data-generated products and the uses and applications of these products that would maximize their value. Prior to focusing on these pertinent issues, however, we will further pursue the circumstances that have led and continue to contribute to the plight of aquatic remote sensing. This plight, in essence, defines the arena in which veteran and rookie proponents of the value of remotely acquired case 2 water quality products must perform if this value is to be articulated convincingly and thereby incorporated within the protocols of environmental monitoring.

1.4   The irony of remote sensing

To be an effective tool for environmental monitoring, remote sensing must have applicability to one or more of the following directives: to locate; to identify; to demarcate spatial and temporal change; and to assist in the assessment of such change. In addition, remotely sensed data must be readily available to users in a reliable form that can be manipulated mathematically. Although *unambiguous* location, identification, and demarcation of spatial and temporal change are of paramount importance, often *less than unambiguous* location, identification, and spatial and temporal change can be significantly valuable to environmental assessments. (We shall revisit this concept in Section 1.9, which deals with an inevitable remote sensing compromise.)

Environmental satellite missions are designed to acquire information on the *biological* status of Earth's ecosystems. No instruments currently exist that can directly measure biological variables from space. Because remote sensing devices record *physical* electromagnetic emanations from some combination of organic and/or inorganic matter comprising the biosphere, these electromagnetic data must be converted through appropriate multidisciplinary modeling activities into inferred estimates of the environmental variables sought. Therefore, development of *biophysical* models and algorithms is essential to remote sensing.

Most satellite data models and algorithms require knowledge of the optical properties of the organic and inorganic biospheric matter admixture indigenous to the local environmental target or scene. Invariably, such admixtures are temporally and spatially variable. Consequently, the development of general or universal environmental parameter-extraction algorithms for remotely sensed data on a global scale is highly unlikely. Extraction algorithms, therefore, are constrained to be local in application (particularly if such algorithms are based on regressions between ground-measured environmental variables and satellite-recorded values of radiation at some wavelength or group of mathematically entwined radiations from several wavelengths). It is possible to generate theoretical models that possess degrees of acceptable rigor and robustness. However, application of these models invariably requires precise numerical values of the optical properties of the indigenous ecosystem membership and/or the rates of photobiological and photochemical productivity as a function of wavelength (action spectra) pertinent to the ecosystem under study. The ubiquitous presence of and dependence on models will be an obvious subtheme throughout this book.

Remote sensing is excellent at providing a time series of changing color and changing temperature. Changes in environmental color and temperatures (two principal parameters contained within the impressive archives of satellite data collected since the launch of ERTS-1) are consequences of environmental stressors. Environmental stress can result from

- Some combination of changes in composition or physical–chemical properties of the atmosphere
- Introduction of unexpected exotic species into the existing memberships of ecosystems that have established a dynamic equilibrium
- Anthropogenic disruptions to the carbon, nitrogen, oxygen, inert gas, or other chemical cycles of the environment
- Recurring, although nonperiodic, step-function climate events of varying magnitude and location

Changes in environmental color may also be indicative of "natural" progressive evolutions of a nonstressed ecosystem. Therefore, remotely sensed data are spatial "snapshots" of the cumulative, synergistic, and integrated consequences of *all* the environmental stressors (anthropogenic and natural) governing the behavior of the ecosystem (terrestrial, aquatic, combined) at a given moment in time. Reducing a resultant environmental color into conjoint cause/effect relationships responsible for that color can be quite formidable and fraught with uncertainty because it is conceivable that the same environmental color could be a consequence of a number of totally different combinations of environmental composition and/or environmental stresses. The interdependence of terrestrial and aquatic ecosystems illustrates the importance of not considering remote monitoring of inland and coastal aquatic systems in isolation from their adjacent or encapsulating land masses. For environmental monitoring of case 2 water quality, therefore, we would recommend a monitoring unit no smaller than a watershed, basin, or substantial stretch of land/water ocean coast. Geographic diversities dictate that the seamless linking of such monitored units, where and if possible, would require additional consideration.

It is indeed ironic that the greatest virtue of remote sensing — its ability to provide time-series records of the integrated impacts of environmental stress — provides its greatest challenges — the need to develop and apply biological, chemical, and geological models to extract estimates of nonphysical variables from copious streams of spectro-optical data, as well as the further need to develop and apply multivariate analyses to deconvolve, when necessary, the impacts of specific stressors from the cumulative and synergistic responses of the ecosystem membership to the totality of stressors. That is, the greatest strength of remote sensing also serves as its Achilles' heel.

The successes or failures of satellites in evaluating environmental impacts, therefore, clearly depend upon the models and algorithms developed and used for the extraction of environmental information from the continuum of spectro-optical data collected by Earth-orbiting satellites. It is most unfortunate that, apart from occasional obvious failures and successes, no definitive consensus exists within the remote sensing community as to the relative merits and shortcomings of competing models and algorithms. Calibration and validation of existing and developing models and algorithms remain a major source of frustration for remote sensing scientists. It is perhaps even more unfortunate that newer, more sophisticated satellite and sensor systems are scheduled for imminent launchings while vast archives of remotely sensed data already captured from a galaxy of Earth-orbiting satellites (such as ERTS/Landsat and Systeme Probatoire d'Observation de la Terre, SPOT) since 1972 lie virtually unanalyzed, possible consequences of (among others) the prohibitive costs of obtaining time-series satellite data and technology push continuing to lead end-user pull. Not unexpectedly, the sophistication of radiometric sensor space hardware and tracking station data-capture software exceeds the sophistication of existing data interpretation methodologies.

1.5   End-users of aquatic remote sensing products

End-users of remotely sensed data can be divided into four broad categories. Although the following categories focus on interests and activities of end-users of case 2 water quality products, the four broad categories apply to all end-users of remotely acquired data, whether the data are aquatic or terrestrial.

1.5.1   The remote sensing scientific community

Members of this community

- Conduct laboratory and field studies of the optical properties of inland and coastal waters
- Develop and modify theoretical models to explain the nature of in-water radiation fields and the spectro-optical interactions that accompany the propagation of radiation through the water column
- Use these models to devise algorithms to extract water quality information from the upwelling radiance distribution recorded at a remote platform
- Use this information to effect long-term studies of aquatic change

These scientific end-users develop the remote sensing of case 2 waters, are responsible for its successes and failures, and may be found in university, private sector, and government research institutes or departments concerned with the science of remote sensing. Included within this group are educators who use the methodologies, reports, and textbooks generated by spectro-optical research as instructional material for students. Thus, the remote sensing scientific community becomes its own end-user.

1.5.2 The nonremote sensing scientific community

Members of this community use but do not actively do, understand, or necessarily show interest in remote sensing *per se*. These scientists represent a manifold of disciplines that investigate part or parts of an aquatic ecosystem and may be found in university, private sector, or government research laboratories. Their research may vary from molecular to integrated environmental scales. Too often, however, they are unaware of the advantages (or possible lack of them) that aquatic remote sensing may bring to their research interests. The simplest and most straightforward illustration of their unawareness of the value of remote sensing is that the vast majority attempt to understand and record spatial variability of a lake or lake-bottom constituent by collecting massive amounts of samples at different locations. They make several explicit and implicit assumptions and extrapolate this information to a spatial scale that is a million or even a billion times larger. Even more damaging (and trivializing) to the plight of aquatic remote sensing has been this end-user community's historically unstoppable practice of using a satellite image to provide the cover of a report, thesis, or publication generated without involving remote sensing as anything other than an aerial camera to provide a conveniently impressive "high-tech" picture of the test site.

1.5.3 The private sector community

Members of this community are end-users who convert remote sensing products into profit. Included are vendors who market GIS (geographic information systems)-compatible environmental decision-making computer systems; acquirers of airborne remotely sensed spectro-optical data over environmental targets; environmental consultants; and manufacturers of above- and below-water instrumentation for case 1 and case 2 water research. Invariably their main client is also the scientific user group (remote sensing and nonremote sensing), so once again the scientific community becomes its own principal end-user.

1.5.4 The decision- and policy-making communities

Members of these communities utilize (or ought to) the results of the scientific community to manage, direct, and regulate public and private sector activity and behavior through establishment of guidelines, policies, and laws. Also, the policy-making community provides public funds to finance scientific research and student education. These decision-makers and policy-makers are invariably politicians or public officials ranging from local to national to international government. Decision-makers and policy-makers would logically comprise those charged with implementing measures of sound environmental stewardship. These end-users are not as yet responding to the technology push; as we all know, pushing against immovable objects accomplishes zero work.

The complexity of this end-user community is further illustrated by the tenure of membership. In addition to decision-makers and policy-makers, this end-user community also includes the executors of decisions and policies. A sense of permanence is associated with these executors (as with scientific end-user groups), but a sense of transience is associated with those charged with decision-making and policy-making. Here we encounter an essential aspect of this introspection: policy-makers make policy based (hopefully) on relevant information of the physical and socioeconomic environments. For remotely derived information to be part of this policy-making process, it is necessary to transform remote sensing data into spatial information and then to transform this information into information relevant to policy-making.

This necessitates reducing a sophisticated spatial thematic map to a simple managerial-directed statement such as, "On a yearly basis our waters are above or below the threshold of eutrophication" — irrespective of the merit of that threshold, or "The remedial actions taken to reduce the outfall from that nuclear power facility are working" — irrespective of the merit of a preselected threshold. (This situation is exemplary of discussions dealing with the inevitable remote sensing compromise in Section 1.8.)

1.6   Aquatic remote sensing products

Principal remote sensing water quality products result from the values of the coextant concentrations of organic and inorganic colorants present in the upper attenuation length (penetration depth) of a remotely monitored water body (See Section 5.5). If the bottom substrate is contained within this upper attenuation length, information on the nature of this substrate and/or its biological cover would also qualify as a water quality product. Thus, water quality "maps" include, on a per-pixel basis, actual values (in best-case scenarios) or relative values (in minimum-acceptance-case scenarios) of coexisting concentrations of chlorophyll, suspended minerals, and dissolved organic carbon. The concentrations of suspended organic matter (SOM) would also be required water quality products for shallow eutrophic lakes or lakes in peat-rich areas where organic material may comprise over 70% of the suspended matter.

For convenience, aquatic remote sensing products are considered as stratified into four levels, with each level representing the successful completion of an increasingly complex task in the conversion of data into information.

Level 1 products (geophysical):

- Raw, unprocessed spectro-optical data collected and recorded at the remote sensing platform

- Raw, geometrically registered spectro-optical data collected and recorded at the remote sensing platform
- Geometrically registered spectro-optical data that have undergone some correction for atmospheric intervention and/or sun-glitter effects

Invariably, the end-users of level 1 products comprise the remote sensing scientific community, which uses these data to calibrate, validate, or modify existing models and algorithms, develop new ones, and/or generate level 2 products. The first, and arguably the major, source of uncertainties that will cascade throughout the generation of subsequent higher level products occurs at this level. Various models have attempted to extract the unwanted atmospheric signal from the radiance spectrum recorded by the satellite sensor (in the case of water color, this unwanted signal is essentially the color of the intervening atmosphere). These models have ranged from the overly simple attribution of all return in a single wavelength (or wavelength band) to the atmosphere to diligent generation of "generic" atmospheres from practical (although hypothetical) combinations of atmospheric type, air temperature and humidity profiles, solar zenith and azimuth, aerosol type and density, time of year, and a host of other optional and obligatory atmospheric parameters.

Despite individual user preference, the intractable patchiness of atmospheric aerosols ensures that no one atmospheric correction technique clearly eclipses the others in eliminating atmospheric obfuscation. Thus, successful remote sensing ventures often become very local in the sense that site measurements of atmospheric conditions and properties become essential components of the venture. Sadly, a significant portion of the remote sensing community adopts a rather cavalier attitude towards this atmospheric correction problem, almost hoping that a self-fulfilling prophecy could be invoked (namely, if the atmosphere is ignored, it will go away). This cavalier attitude is not necessarily all bad, of course, because this desire on the part of water researchers to avoid the atmosphere has resulted in the development and implementation of ship-boom-mounted upwelling and downwelling radiometer systems that contend with less than a couple of meters of atmosphere below the sensor configuration. One such duo-radiometer system was the ship-boom-mounted CZCS mimic of Bukata et al. (1988a).

Although research into atmospheric correction is an ongoing activity (e.g., use of water vapor absorption bands in the 900- to 1000-nm airborne data to estimate the concentration of water vapor in the atmospheric column beneath the sensor), this potential obstacle to remote sensing of case 2 water quality remains to be satisfactorily overcome. Wind-related weather patterns follow generally predictable west-to-east directions across much of the Earth. Coastal zones on western sides of continental land masses, therefore, would generally be characterized by narrower bands of case 2 water regimes than coastal zones on eastern sides of continental land masses; this is a consequence of wind-driven terrestrial runoff being considerably more pronounced from eastern shorelines.

A comparable condition exists for an atmosphere that contends with a wind field that moves from west to east across a continental land mass. Consequently, although aerosols comprising west coast marine atmospheres pose severe problems to remote sensing, they are less problematical than aerosols comprising east coast atmospheres. The additional optical complexities of the atmosphere and the water conspire to make atmospheric correction more difficult for eastern coastal remote sensing. The converse is true for those regions of the Earth where wind-related patterns follow generally predictable east-to-west patterns where optical complexities of atmosphere and water are considerably more pronounced for west coast marine atmospheres than for east coast marine atmospheres. In either case, atmospheric correction for inland and coastal waters is generally the most formidable.

Level 2 products (bio–geo-chemical):

- Geophysical variables such as water-leaving radiance, attenuation coefficient, and remote sensing reflectance at the air–water interface
- Estimates of concentrations of aquatic color-producing agents (CPAs) such as phytoplanktonic pigments, suspended particulate matter, dissolved organic matter, and suspended organic matter
- Locations of submerged surfaces such as shallow bottoms and coral reefs

The end-users of level 2 products once again comprise the scientific community that uses these geophysical, biophysical, and CPA data to calibrate, validate, and modify existing models and methodologies and formulate new ones to generate level 3 products. These models and methodologies are the mainstays in converting water color into the concentrations of the CPAs collectively determining that color. Thus, level 2 products are the first big steps towards inland and coastal water quality monitoring.

Level 3 products (aggregation level 1):

- Indices for primary production, bioproductivity, eutrophication, and/or pollution

The end-users of level 2 products comprise the scientific community that uses this information (notice that remote sensing data have been converted into information by level 3 — thus, the aggregation level 1 subappellation) to calibrate, validate, and modify existing models and methodologies and formulate new ones to generate level 4 products. However, the scientific end-users are joined by a segment of the water resource management community that utilizes such indices to determine (or at least to illustrate) that certain regions of the environment are at risk; whether certain regions of the environment should be subject to rehabilitation; or whether certain rehabilitation measures already in place are satisfactory. Unfortunately, only a limited number of such indices exist and those that do are still being evaluated for possible use in "state-of-the-environment" reporting schemes. Their use is still debatable and certainly far from operational.

Level 4 products (aggregation level 2):

- Regional-scale (watershed or coastal) state-of-the-environment (ecosystem) assessment
- Large-scale (linking of watersheds or extensions of coasts) state-of-the-environment (ecosystem) assessment

Although the end-users of level 4 products still include scientists, the targeted end-users of these products are water resource managers, environmental decision-makers, and political policy-makers. Sadly, however, level 4 products are currently nonexistent or in early stages of definition and development (in the hands of the technical and scientific research community that is chaperoning the level 1 to level 4 transformations). Production of reliable level 4 products by the research community and their acceptance and use by political policy-makers essentially control the fate of remote sensing of inland and coastal water quality. It is essential that these level 4 products be well defined and demonstrably valuable. However, beyond the generation of level 4 products, the influence of the scientific community rapidly diminishes, if indeed it does not vanish completely.

Clearly, major developments are still required for the generation of level 3 and level 4 products (in addition to the need to solve the atmospheric intervention difficulties plaguing the generation of level 2 products). Only part of these developments resides within the province of logistics and technologies of satellite orbits and sensors (better spatial and temporal resolution, more frequent site visitation intervals, real-time data acquisition, telemetry, and so on). The bulk of these developments resides within the province of data interpretation, conversion of data into information, and presentation of information to influential end-users outside the scientific community.

1.7   What went wrong with the great potential?

Remote sensing of the Earth's environment from space vehicles was born in the socially, morally, and politically turbulent 1960s. Developed and some developing nations were struggling to replace a long-standing complacency characterized by a confounding combination of real and perceived ills and inequalities with an idyllic lifestyle of universal peace, beauty, and freedom and a global inclusiveness of mankind in harmony with all of Nature's creatures and resources. Consequently, the 1960s quickly became rife with the philosophical advocacies of reformers, lobbyists, protesters, students, poets, actors, and folk singers.

Music is a powerful, influential force for change, and the airwaves were filled with conscience-nudging lyrics. Two of the most powerful and influential songs were the prophetic 1964 Bob Dylan composition "The Times They Are A-Changin'" and the 1965 Chet Powers counterculture anthem "Get Together." The former was a stern warning to politicians and policy-makers that they must quickly adapt to new and impending all-inclusive social structures that would not be denied, and the latter was a directive that these impending social structures be free spirited and comprise requited and universal love. These musical directives combined with other songs, including

- The poignant lyrics of Paul Simon's "The Sound of Silence"
- Quasireligious rock hymns such as Simon's "Bridge over Troubled Waters," John Lennon and Paul McCartney's "Let It Be" and "Hey Jude," and the entire score of Andrew Lloyd Webber's rock opera, "Jesus Christ Superstar"
- Gerome Ragni and James Rado's tribal rock musical "Hair"
- Civil rights anthems such as Guy and Candy Carawan's adaptation of "We Shall Overcome," Tom Paxton's "What Did You Learn in School Today?" and Alan Robert-Earl Robinson's "Black and White"
- Antiwar songs such as Pete Seeger's "Where Have All the Flowers Gone?," Dennis Lambert and Brian Potter's morality fable "One Tin Soldier," and Buffy Sainte-Marie's "Universal Soldier"
- Scenarios of global nuclear holocaust such as Phil Sloan's "Eve of Destruction" and Fred Hellerman and Fran Markoff's "Come Away, Melinda"
- Songs in favor of restoration of the environment such as Joni Mitchell's "Big Yellow Taxi," and Albert Hammond and Mike Hazelwood's "Down By the River"
- Peter Yarrow and Leonard Lipton's debatably metaphoric "Puff (the Magic Dragon)"
- George Weiss and Bob Thiele's wistfully plaintive "What a Wonderful World" (immortalized by jazz legend Louis Armstrong)

Along with countless other lyrics expounding and modulating the troubled backdrop of changing values and agendas, these songs serve to illustrate the populist culture into which science and technology, in general, and environmental monitoring from space, in particular, had to be inserted.

The United States was in the final years of its race to the Moon with the (then) United Soviet Socialist Republic. NASA was at the crest of its public supported and popular manned and unmanned space exploration programs. An unpopular and divisive no-win war in Vietnam was being waged, opposed, promoted, and (draft) dodged. Protest groups were continually staging marches and sit-ins for a broad spectrum of causes ranging from poverty to civil rights to legalization of mind-expanding drugs.

The Earth's environment quickly joined this spectrum of causes. The "future shock" vision of environmental destruction by human hand became entrenched in public and political thought — a vision that, although radical, was not without verifiable justification. Unfettered pollution was rampant, lakes were becoming eutrophic, forests were in decline, strip mining was taking its toll, and the local/global environment and the creatures that it harbored were being threatened. NASA anticipated let down of public support once lunar landing became a reality and therefore recognized a need to display relevance to the mind-sets of North Americans. Reaching Mars seemed like science fiction. Thus, physical studies of near-solar and inter-planetary media from space were replaced overnight by biological studies of the Earth's environment from suites of satellite sensors looking toward the Earth rather than spaceward. As fate would have it and history was to show, difficulties to be encountered in this transference from physics to biology were very much underestimated.

In addition to adding a much needed relevance card to NASA's deck, through research and its ensuing applications, ERTS-1/Landsat-1 was to provide an ability to monitor environmental behavior and ecosystem health synoptically from space. Despite its being an unproven asset of an as yet untried venture, ERTS-1/Landsat-1 was heralded and promoted by its overly zealous supporters as a program of virtually unlimited "great potential." To facilitate this program and "kick start" its potential, NASA prepared to provide these synoptic data readily in photographic imagery format to a wide and varied environmentally minded clientele. This group included government, university, and private sector workers in Earth-resource areas, as well as self-declared environmental activists whose substitute for scientific and academic credentials was emotion, a sense of righteousness, and missionary zeal. In fairness, however, it must be realized that NASA was not alone in such zealous widespread promotion of environmental monitoring from space. Government agencies within Canada, Europe, Asia, India, and other world centers quickly and happily followed NASA's lead in national and international satellite programs. Such overselling of products not yet conceived, however, would result in a "buyer beware" label for them once they were conceived.

Although this chapter "navel gazes" on inland and coastal water quality monitoring, much of the introspection and retrospection herein can apply to terrestrial ecosystem monitoring as well. As discussed in Section 1.5 and Section 1.6, the dedicated research community comprises two of the four categories of end-user groups, as well as the end-usership of at least three of the four level products resulting from the remote sensing of case 2 waters. However, this should not paint the aquatic optics research community as an incestuous self-serving conclave. Such a situation is not specific to aquatic optics. It is an historical problem experienced by virtually every scientific discipline or activity — in particular, new, novel disciplines or activities lacking an extended history and an established respected pedigree. Nonetheless, evidence of the relevance of remotely monitoring the quality of inland and coastal waters must be expanded beyond the scientific community that is making such monitoring possible. We are currently knee-deep in

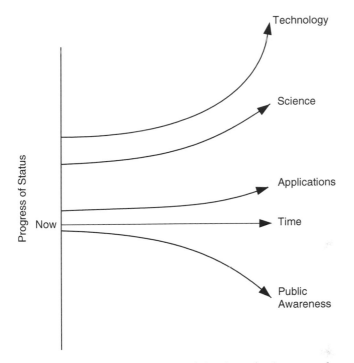

Figure 1.1 A simple, schematic illustration of the disparity between the promise of remote sensing and the actual delivery of the promise to end-users, based upon suggestive differentials in the rates at which progresses in technology, science, applications, and public awareness are currently proceeding.

a quagmire of resounding influential end-user disinterest. Decision- and policy-making end-users must buy into the relevance; however, to date, this has not happened (although, as we shall discuss in Chapter Six, a number of recent signs should encourage guarded optimism).

Figure 1.1 is a suggested (although exceedingly simple) schematic illustration of the disparity between the promise of remote sensing and its actual delivery to end-users. In essence, it represents the ongoing consequences of technology leading science, science leading applications, applications leading end-user awareness, and end-user awareness leading end-user interest. Left unchecked, an ever widening gap is generated between technology and end-user interest. That this creates an even more critical gap is evident when one considers the major players in this game of catch-up. Remote sensing scientists and technologists comprise (say, very generously) considerably less than 1% of the population of a developed country. Space monitoring is an expensive activity for less than 1% of the population to be engaging in a game of catch-up that will never be won (particularly when 100% of the employed populace is taxed to pay for it).

NASA realized this need to involve the public in the 1960s when it printed the relevance card for its unmanned space program. To the general public, however, this relevance card was not a "collectible card." Thus, as technical advances in space hardware and software were made, the general public (passive sponsors of the catch-up game) was essentially excluded; therefore, its knowledge of and interest in remote sensing applications started low and has rapidly diminished. Therefore, as technology, science, and applications are following exponentially *increasing* time lines (although with differing exponents), public knowledge, awareness, and interest are following an exponentially *decreasing* time line. This rapidly increasing gap between the promoters and the public sponsors of satellite monitoring of Earth's ecosystems could very easily bring such monitoring to an abrupt halt when other demands are placed upon the tax dollar (particularly during periods of faltering economy). It is therefore almost trite to say that Figure 1.1 illustrates the need to

- Close the gap between technology and end-user interest by perhaps redistributing some of the less than 1% catch-up game players from the leading edge to the trailing edge of the gap.
- Present compelling arguments to potential end-users of space-acquired environmental data products.
- Perhaps most important of all, educate, inform, and convince the sponsoring public that the catch-up game is being and ultimately will be won and that the sponsoring public will be the benefactors of the win.

Thus, public schools, universities, and junior colleges have major roles to play — as do the public media — and all must be brought on side. Of course, all this becomes academic if the sizes of the gaps do not reduce.

Although there is an encouraging slowly rising trend of end-user involvement in Earth observations from space, most of this interest is in terrestrial ecosystems (farming, forestry, land use) and mid-ocean dynamics. (Weather, which has the principal satellite end-user interest and will be discussed in Chapter Four, is not an environmental ecosystem activity *per se*.) Unfortunately, end-user disinterest in inland and coastal aquatic ecosystem and water quality monitoring products remains widespread and relatively steadfast. A partial list of the reasons for this relentless end-user disinterest might include:

- The technology is inadequate.
- The science is inadequate.
- The applications are inadequate or nonexistent.
- The communication of potential, results, and products is inadequate.
- No official standards for remote sensing products exist.
- The costs are too high for the deliverables.
- Remote sensing of the Earth from space is generally suspect.
- Societal and/or organizational barriers exist.

Why, when, and how did the great potential go astray? Let us briefly examine each of the preceding issues.

1.7.1   *The technology is inadequate*

As discussed in Section 1.1, sensor technology appears to be quite adequate, especially if the combined capacity of air- and space-borne sensors is considered. In general, estimation of water quality through remote measurements of water color requires high spectral and spatial sensitivity from space-borne imaging spectroradiometers. Although increased spatial resolution is needed and would be most welcome, current and planned satellite sensors actually have spectral sensitivities in excess of what is needed or even manageable for most water quality determinations. Thus, from the perspective of water color, upcoming hyperspectral satellite sensors present exciting opportunities for generating inland and coastal water quality products.

Categorizing software under technology might be opening a Pandora's box of controversy. Is the processing software good enough? The answer is probably not. We are not aware of any turnkey software solutions that take a remote sensing image and, at the push of a button, create a reliable chlorophyll map of an inland or coastal water body. The MODIS and SeaWiFS standardized software does produce state-of-the-art maps of chlorophyll and other water variables for open-ocean waters. However, the accuracy, and thus the environmental value, of the MODIS coastal products is not obvious, so it is open to debate. Also, the scale of 1-km pixel size is too coarse for most inland and coastal water quality applications. Granted, coarseness is a hardware consideration, although the software is confined by this coarseness as well.

Another technological issue is the inability to see visible radiation returns through atmospheric cloud or haze (to be discussed in greater detail in Chapter Two and Chapter Three). This results in an unfortunate dependence on so-called "cloud-free" scenes. The preponderance of atmospheric clouds over inland waters (e.g., Laurentian Great Lakes of central North America) is a frustratingly common reason why more optical satellite data are rejected than considered for water quality analyses. Perhaps too little research has been done to determine the minimal circumstances under which relevant information can still be gleaned from a remote sensing image (e.g., is 25% cloud coverage a problem or not? Can a roughly homogeneous overcast be treated as an end-point of a cloudless clime? Under what circumstances can radiance data recorded from sporadic "breaks" within extensive cloud cover be of consequence?).

1.7.2   *The science is inadequate*

It would indeed be presumptuous as well as inaccurate of us to suggest that remote sensing science is beyond reproach. It is, nonetheless, very good. This very good, although nonoptimal, status of bio–geo-optical models along with their impacts on water quality products will be discussed in Chapter Two and Chapter Three.

1.7.3   The applications are inadequate or nonexistent

Applications of satellite monitoring of water color to environmental monitoring and decision- and policy-making most definitely exist. These applications form the impetus for this book. Chapter Four will deal with applications of space monitoring of inland and coastal water quality to environmental stewardship by addressing the aquatic and terrestrial priorities contained within the Environment Canada Nature Research Agenda (Environment Canada, 1999a, b).

1.7.4   The communication of potential, results, and products is inadequate

Irrespective of the position from which we view this issue, the inescapable observation is that as a remote sensing development community, we have not communicated well and have not necessarily shown the right results or developed the right products. Undoubtedly, the pride and comfort accompanying the admittedly excessive communication among technical and scientific peer groups (to the almost total exclusion of all other groups) became an inadvertently constructed barrier preventing our knowledge and capabilities from becoming widely known outside the fields of aquatic optics and remote sensing.

However, it is most likely that, even if we had not consistently excluded "nonclub members" from workshops, scientific gatherings, and project planning sessions, we would have failed to communicate to them at the level of applicability that potential end-users were seeking. Irrespective of other enviable talents, scientists (especially government and university scientists) are notoriously ineffective at marketing their science to anyone outside the science field. That the proponents of remote sensing have not properly spoken to the needs of potential end-users of water quality products is evident. These needs may not have been adequately addressed in time, quality, quantity (e.g., as a series in time), price, or indeed in all of these areas.

It is also evident that the water quality products, although convenient for customers, are not customer friendly. This is often a consequence of too much fine-tuning by the data specialists involved in generation of the product. Invariably, end-users require spatial information (usually a one-off of a specific site or area, but occasionally a time series of spatial information). Managers and their technical teams are becoming adept at handling GISs and ISSs (information support systems). However, in the majority of instances, the information given to the manager is the output of a purchased computer program involving a model or string of models (possibly less than properly mathematically interlinked) with which neither the manager nor the technical team is fully cognizant. The purchased computer program is conveniently regarded as appropriate to the environmental issue at hand and its outputs considered accurate.

The relative ease with which such computer programs may be obtained has blurred the demarcation between "modelers" and "purchasers" and, as such, is systematically and rapidly replacing modelers with program purchasers. As discussed in Section 1.8, reducing the relevancy of scientific modelers in the application of their modeling efforts is not advancing the acceptance of space-acquired inland and coastal water quality products. Proliferation of such readily available packaged computer programs has probably been responsible for the generation of inappropriate information upon which inappropriate decisions have been based. Obviously, this has been detrimental to the plight of remote sensing from space vehicles.

1.7.5   No official set of standards for remote sensing products exists

No national institute of standards has ever established standards for remote sensing products. Historically, remote sensing has an abysmal record when presented in court as evidence. In fact, because of the professional embarrassment routinely and gleefully inflicted on remote sensing "expert" witnesses by lawyers, experts are understandably reluctant to present such evidence. This reluctance is abating somewhat, particularly in the instances of oil spills in which an image can often convincingly associate the spill with an unequivocal source (e.g., ship, oil rig, industrial complex). Reluctance still exists when remote sensing is presented as evidence for the magnitude of the spill (volume and/or impact). Compliance with a national or an international water quality protocol has yet to be verified by the use of remote sensing. The Canadian–American International Joint Commission (IJC) has used satellite data in transboundary transport issues, but no definitive actions or policies of consequence have ensued.

Many water quality management authorities (WQMAs) on a global basis have large monitoring budgets but small research budgets. Often, national, state, or provincial laws compel them to use methods based on field sampling or field measurements despite the fact that remote sensing could produce a better result for a specific required task or series of tasks.

1.7.6   The costs are too high for the deliverables

The answer to the title's implied question is yes and no. Remote sensing data acquisition costs of purchase of acquired satellite data or collection of new airborne data are often seen as a barrier for applications of remote sensing for management agencies with inland, coastal, or marine mandates. The problem is that, despite continual contrary pronouncements by the remote sensing community, remote sensing is often considered to be competing with established ground-based monitoring protocols. This "apples and oranges" cost comparison should perhaps be considered in a manner in which the local currency becomes "cost per monitored area." Under such currency, established techniques for monitoring water quality variables (water samples collected from a network of fixed stations or transects) may be more expensive and less spatially inclusive than those for remote sensing techniques.

Without becoming embroiled in discussions of actual costs of ship and aircraft deployment, field logistics, laboratory facilities, computer facilities, and labor (which vary significantly with geographic location), broad statements regarding the "cost per monitored area" currency approach to evaluating cost/deliverables might be reasonably made. For example, the answer to the implied question could be no (i.e., the cost is not too high for the deliverables) if a satellite image can alleviate the need to obtain a water sample every square kilometer of a monitored water body — i.e., when the cost of satellite sensing is compared with the cost of one water sample per square kilometer. However, the answer is yes if the cost of satellite sensing is compared with the price of one water sample per 1000 km^2. The answer is also yes if satellite sensing does not replace or substantially enhance any existing technique as a remote sensing-derived product because remote sensing then becomes an additional cost. As a general rule, satellite data purchase and processing costs are often lower per unit area (as opposed to airborne data acquisition and processing) because of lower data volume per area. However, differences in spatial resolution must be factored into this comparison.

Often, remote sensing products can replace or dramatically improve part of the monitoring program of a WQMA (or of a lake management authority, LMA). However, the WQMA or LMA would still need to maintain the infrastructure for its other ground-based measurements (e.g., it would no longer need to measure for chlorophyll and total suspended particulates, but it would still need to sample for organic micropollutants and heavy metals).

Obtaining the three decades worth of time-series satellite data (Landsat, SPOT, etc.) is a luxury outside the financial envelopes of most research scientists. Despite its restricted spectral resolution, this grand archive of space data would support multidisciplinary impact-related research in many ecological areas. Such a valuable historical record of changing states of regions of the global environment would provide tempting glimpses of the individual, collective, additive, and synergistic responses of environmental variables and their suite of environmental stressors. Revising the methods of sale of historical and current satellite imagery to the user community is urgently needed if its potential for applications to local and global changes in ecosystem health is to be realized.

Satellite data are currently sold on a per-scene or per-strip-of-scenes basis. Bukata and Jerome (1998) suggested that satellite data be sold by the bit rather than by the scene, illustrating that the Landsat data (irrespective of cloud cover) collected over a selected test site of 45.1 km^2 over a 26-year period could be purchased for the current cost of one full-scene image. To date such a suggestion has been ignored by satellite data vendors. However, suppliers of recent satellite data have made welcome concessions to the user community. Recent selected satellite data and their level 1 and level 2 products may be downloaded from excellent World Wide Web sites; some of these data may be edited. Also, some data vendors have occasionally been receptive to discounting archival satellite data as long as a substantial enough amount of time-series data is purchased.

1.7.7   Remote sensing of the Earth from space is generally suspect

The suspect nature of remotely sensing Earth from space is an unfortunate reality that must be reluctantly borne. It may be the largest cross that remote sensing has had to bear since its irresponsible promotion by well-meaning, although misguided, proponents as a potential cure for environmental ills yet to be uncovered. In an attempt to display relevance to the mind-sets of the 1960s and 1970s, NASA played to the "all inclusiveness" of these mind-sets. "All inclusiveness" is the antithesis of "elitism"; although certain social cultures *might* attain a degree of self-sustaining all inclusiveness eventually, the science culture *cannot* because of its creation by and evolution through elitism.

All inclusiveness ultimately results in good becoming indistinguishable from bad. Under inbred doctrines of all inclusiveness, science atrophies and eventually becomes at best a diversion or at worst a pseudoreligion proselytizing prevalent political correctness. In transforming space science into environmental remote sensing, NASA also circumvented its hitherto stringent and successful rules of participation. Its manned program of lunar exploration and its unmanned program of solar and interplanetary medium exploration attained such highly successful results in such a short space of time because it wisely put them into the charge of scientific and technical elitism. In the unmanned exploration of space, the elite scientific community competed for a limited number of "seats" on each Earth-orbiting satellite and/or deep-space probe. Successful "passengers" then installed their conceived, designed, and fabricated sensors into those seats and developed all the required data capture, storage, and analyses schemes to transform those data into scientific knowledge.

Furthermore, they all had well-defined scientific goals and execution plans that had been severely scrutinized by peers, including mandatory defense before the congressional space science steering committee. Most importantly, successful passengers understood the roles of their specific instruments in achieving their goals. Also, the passengers were usually given a 2-year monopoly on their acquired space data prior to the data entering the public domain. None of this, however, was the case for satellite monitoring of the Earth's environment. There were no competitively successful passengers. Prior to launch, multispectral scanning sensors had been designed by a small core of scientists and engineers assigned the Herculean task of making those sensors applicable to the diverse and ill-defined needs of hosts of environmentally concerned potential end-users with little to no background or interest in space science.

Involved from the start to the end of every unmanned mission, the elite space scientists produced elite space science. However, space science is space physics, and elitism had already established physics as an undeniably credible discipline. Remote sensing was multidisciplinary in makeup, was not an established credible entity, and, of course, lacked an essential elitism. Data from ERTS-1/Landsat-1 were made available in photographic image format to a widespread, open, and diverse group of potential users — the majority of whom knew little to nothing about the virtues and/or limitations of remote sensing or the content of the data collected by remote sensing devices.

The elite of each individual discipline comprising the disciplinary sciences (physics, chemistry, biology, geology) and various multidisciplinary sciences (environmental optics, oceanography, forestry, limnology), which were essential if physical parameters were to be believably converted into estimates of ecosystem health, were buried within the overwhelming hordes of all inclusive owners of pictures. While the elite was developing multidisciplinary models and patiently waiting for some digital data to analyze, the picture owners, who readily recognized major cities, mountains, and summer cottages, became instant experts in something nebulously referred to as remote sensing. Elitist science was shunted to the sidelines in favor of a quasisurreal pseudoscience generated by the instant experts. Such quasisurrealism appealed to the fabulists. It was as though the alchemists' 500-year quest for the philosophers' stone had come to an end. The stone was in a sun-synchronous, near-polar orbit about the Earth. Could the elixir of youth, the universal solvent, and perpetual motion be far behind?

Volumes of totally inappropriate (and, even worse, erroneous) success story literature (most in the so-called "gray" literature) were generated by these instant experts. Remote sensing had, as yet, neither reputation nor pedigree, so this gray literature, by default, became its reputation and pedigree. The pedigree became readily accepted as genuine and further promulgated by those eager to benefit from global monitoring, thereby fostering understandable and warranted skepticism in end-users who could see the forests *and* the trees. The scientific community was understandably reluctant to embrace environmental remote sensing seriously.

To add to the woeful status of remote sensing and possibly also as an overture to the all-inclusive mind-sets of the era, Harvard Business School introduced its *management by objectives* concept, which was immediately adopted by educational institutes and government agencies across North America. To Harvard Business School's credit, however, it shortly withdrew this concept from its curriculum, admitting it to have been an open invitation to mediocrity. Perhaps better late than never, although, sadly, the invitation had already been accepted. To compound the problem further, as the elite were abandoning remote sensing and retreating to the relative security and comfort of their established individual disciplines, the all-inclusive instant experts were planning future remote sensing missions.

Some educational institutions, on a global scale, embraced remote sensing as an upcoming *discipline*, rather than as an upcoming *activity* that could possibly cross-cut already established disciplines. As a consequence, remote sensing (with ill-defined curricula) became prescribed courses in universities and colleges, as well as graduate thesis topics (both generally within geography departments). In some institutes, remote sensing was instantly granted departmental status; its principal students were aspiring geographers and, at least initially, a significant number of aspiring agriculturalists. Some institutes considered granting degrees in remote sensing and, for a short time, a small number actually did. Although there is most certainly a contingent of elite geographers, such elitism could not develop a respected pedigree for space monitoring of environmental ecosystem behavior on its own.

Thus, with the academic concept of remote sensing being somehow an off shoot discipline of geography, the scientific elitism that was outside geography but essential to the development of Earth observations from space was further shunted to the sidelines. A philosophy of "all inclusiveness save for elitism" flourished. If all inclusiveness results in an inability to distinguish good from bad, consider how much more rapidly that inability arrives as elitism is systematically culled from the conglomerate. Although it is tempting to infer a direct relationship between Harvard Business School's ill-fated management by objectives concept and the ill-fated academic concept of a geography-cloned remote sensing discipline, such an inference would probably be a non sequitur.

Remote sensing has had to tote over three decades of inappropriate work and claims as excessive baggage, with many legitimate virtues of remote sensing interred within this baggage. Thus, by forsaking elitism in favor of all inclusiveness, NASA's attempt to make remote sensing relevant almost rendered it totally irrelevant. Inadvertent aspersions, however, should not be cast at NASA or its international counterparts. Aspersions should not be cast at the scientific elitism for abandonment of a child who, through no fault of his own, was born into a turbulent nursery. Against the volatile backdrop of the 1960s and 1970s, the mood and behavior of the public, government, social lobbyists, student body, reform activists, environmentalists, scientists, technologists, and space agencies, although not necessarily practicing conventional wisdom, did make obligatory attempts to adapt to and thereby survive within the new, convoluted and undisciplined, local, and global reality. Within an uninhibited youth-driven Woodstock era, against a libretto of melodic, infectious, mind-bending, politically driven songs by folk artists like Bob Dylan, Pete Seeger, the Weavers, the Chad Mitchell Trio, Joan Baez, Tom Paxton, the Womenfolk, Arlo Guthrie, and Peter, Paul, and Mary, the times truly were "a-changin'." To its credit, NASA did attempt to introduce environmental monitoring from space as a possible nostrum to allay public unrest. Sadly, however, elitism and satellite monitoring of the Earth's environmental ecosystems fell victim to those "a-changin'" times.

The sad combination of elite exclusion and all-but-elite inclusiveness quickly gave rise to an inability to generate space-acquired environmental products that could be convincingly interpreted in a manner that would impress environmental stewards, although the "success stories" continued to abound. Terrestrial space monitoring had the advantage of distinguishable features that could be readily identified in an image. Thus, interpreting changes in land targets displaying permanence appeared more plausible than interpreting changes in water targets displaying transience.

However, due to a limited number of nonoptimal wavelength bands and a severe underestimation of the scientific complexities to be encountered, the conversion of recorded physical spectroradiance into biological status of land features was also problematical. When research encounters a data set that is particularly recalcitrant, the long-standing defendable scientific practice is that the data set becomes a principal suspect in the scientific dilemma. It is, however, never considered the sole suspect. The immediate proliferation of the self-deprecating term *ground truth* (implying aerial falsehood) is evidence that space data sets were most often considered the sole suspects. The tragedy, of course, is that the data sets were and still are far from immaculate. The integrity of the recorded optical data was justly doubted because of such factors as each remote sensing device comprising multisensors that had to be intercalibrated, the line-scanning nature of the remote sensing process, spectral dissimilarities among satellite missions, calibration updates, the intervening atmosphere, and a host of other technical realities.

Thus, a vast series of activities designed to correct the recorded data for pixel registration, orbital adjustments, off-nadir vignetting, sensor aging and motion, cloud cover, and various other demeanors of data integrity was initiated. That such activities are of consequence is irrefutable, and indeed most of such threats to data integrity have been admirably attacked and conquered. However, the single most vital problem — the intractable patchiness of intervening atmospheric aerosols — is still a work in progress and will be discussed in Section 3.7. In addition, when even "corrected" data remain difficult to interpret, the human condition (although inadvertently) calls for doing a good job at what it can do well rather than doing a poor job at what it cannot do well. Because myriad massages can improve the quality of satellite-acquired data, not surprisingly, correcting the radiance spectra recorded at the satellite sensor became an industry unto itself. Data correction, rather than data conversion into information, became the end-product of a large segment of the remote sensing community.

As a grimly blunt demonstration (apart from the manned space race) of NASA's elitist vs. its all-inclusive space policies, consider the unmanned deep-space solar orbiters Pioneer 6 through Pioneer 9 (launched between 1965 and 1969 in tandem with the Apollo missions). From the Pioneer mission (Pioneer 6 became NASA's oldest operational satellite, in solar orbit over 30 years) came the first detailed measurements of solar magnetic field configurations, large-scale solar flares, the solar wind, and cosmic ray propagation within the solar cavity (see, for example, Bartley et al., 1966;

McCracken et al., 1967; Rao et al., 1967; Bukata et al., 1972, as well as many others). A comprehensive understanding of stellar processes and Sun–Earth interactions emerged within about a decade. The Pioneer 6 through 9 program is still touted as one of NASA's most productive and least expensive spacecraft missions in terms of scientific knowledge per dollar spent. Contrast this with the continuing plight of environmental monitoring from space for more than three decades.

Despite obligatory outreach to public educators and news media, in NASA's post-lunar-landing interplanetary and planetary (other than Earth) missions, the organization avoided the all-inclusiveness pitfalls plaguing Earth monitoring from space. Again, it could place the missions in the hands of elite scientists and engineers. Planetary deep-space probes such as Mariners 9 and 10; Pioneers 10 and 11; Pioneer Venus; Mars Global Observer; Mars Global Surveyor; Galileo; NEAR; Mars Polar Lander; Voyager; Viking; and others yielded a wealth of information on planetary and interplanetary processes, as well as recording the physical features of planets, asteroids, comets, galaxies, stars, moons, and atmospheres. However, the missions were space science and thus space physics, wherein the existence of planetary biology and its nonconforming behavior to rigid mathematical formulism was absent from consideration or merely the subject of speculation.

The all-inclusiveness aspect of remote sensing, although weakened, has yet to concede to elitism. Nonetheless, subsequent sound science by combinations of elite "die-hard" and elite young talent dispelled much of the damaging repudiation imposed by early unfounded grandiose success stories. However, despite conscientious and clever work by these dedicated scientists, remnants of this suspect reputation linger.

1.7.8   Societal and organizational barriers exist

Unfortunately societal and organizational barriers are far too prevalent. It is likely that because of organizational barriers, many potential end-users beneath the upper organizational levels have not had the opportunity to assess the reliability of products as components of their operational protocols. Remote sensing has not been incorporated into the sampling protocols of WQMAs or LMAs (other than, perhaps, in a limited number of research studies). Established monitoring protocols possess a familiarity that provides an understandable comfort zone for the directors and executors of aquatic sampling. Thus, it is easy for WQMA laboratories to develop latent suspicion of and dislike for remote sensing data. Such data may be readily seen as threats to operational routines and to work load and job security.

Incorporation of remotely sensed data, they feel, could mean the closure or the reduction of their funded budgets.

Although such threats are usually overestimated, they are nonetheless real. Remote sensing requires additional data analysis skills. It is relatively easy to manipulate tables, bar graphs, and other one-dimensional data. It is less easy to manipulate a thematic image demonstrating that a spatial pattern of a pollutant distribution shows a distinct source and/or a sink. Although remote sensing could possibly replace or improve upon some water quality measurements, on its own, it cannot supply all of the information needs of the WQMA or LMA. This results in an obvious financial barrier because monitoring from remote platforms becomes an additional cost over and above the cost of the conventional protocols.

Often, upper organizational echelons make totally unrealistic demands of remote sensing products that transcend their demands upon products emerging from established technologies. Reasons for this can vary. Perhaps some of these unreasonable demands could be related to the overly zealous and irresponsible sales promotion discussed in Section 1.7.7. More likely, however, they could illustrate a defensive attitude (turf protection?) to maintain a comfortable noncombative status quo devoid of threats to established conventional thinking, operations, and funding.

Another organizational barrier is a natural consequence of the manner in which budgets are distributed among federal, state, or provincial government agencies. Each agency has a distinct legislated mandate. Combined with the competition for public funds, diverse mandates invariably result in reluctance on the part of most government agencies to be involved — alone or in partnership — with remote sensing, which is a costly activity. For remote sensing to contribute to intergovernmental mandate issues, it is essential to assuage this reluctance by seeking all spatial information requirements within and between organizations. Such spatial data could then be considered as shared acquisition data for multiple purposes. Interdepartmental dialogue and collaboration can significantly decrease the cost per indicator per area and thereby decrease the impacts on individual department budgets. Such shared data acquisition costs are currently effectively implemented within collaborations among Environment Canada, Fisheries and Oceans Canada, the Canadian Space Agency, and the private sector regarding inland and coastal water quality monitoring from space vehicles.

1.8   Models, models, models, models

Theoretical models are the framework upon which the remote sensing of inland, coastal, and mid-oceanic waters is constructed. Fundamental principles of in-water optical processes are incorporated within the theory of radiative transfer of energy. All optical models are based upon mathematical simulations of these energy transfers. Optical models and their outputs will be discussed in Chapter Two and Chapter Three. However, as seen from the description of water quality products in Section 1.6, not one level of water quality product may be generated without them,

- Level 1 products require models of the atmospheric absorption and scattering of the solar radiation incident at the top of the atmosphere.
- Level 2 products require forward models, which can generate an aquatic color from the concentrations and optical properties of the indigenous CPAs, and inverse models, which can deconvolve an aquatic color into the coexisting concentrations of the CPAs that have generated that color. These are the models most fundamental to aquatic remote sensing becoming a credible water quality monitoring tool.
- Level 3 products require models that convert estimates of near-surface chlorophyll, suspended inorganic (and possibly organic) particulates, and dissolved organic matter into biologically meaningful descriptors such as primary production (phytoplankton photosynthesis) and consequences of such primary production (e.g., eutrophication).
- Level 4 products require models that, by and large, have yet to be devised because the seamless linking of biological indicators of inland/coastal waters with the biological indicators of adjacent or encapsulating landforms requires the interlocking of land and water models, each genre of which is developed with, at best, token consideration of the other. The variables essential to seamless environmental linkages are terrestrial and aquatic bioproduction and terrestrial and aquatic contributions to carbon cycle dynamics. Because the simultaneous concentrations of chlorophyll and dissolved organic carbon are variables that are estimable from remote sensing, these level 4 products represent an area that warrants the focus of considerable additional scientific modeling. Reliable level 4 products are essential for the acceptance of remote inland and coastal water quality monitoring.

The need for and use of models does not end here, however, even though the active role of the aquatic optics research community usually does. Conversion of remotely sensed data into environmental decision-making and political policy-making transcends the combined talents of teams of multidisciplinary scientific modelers. Environmental policies represent perceived compromises among special-interest groups driven by political, financial, societal, legal, and environmental agenda that are not necessarily in harmony with the science agenda. Thus, once biological variables are inferred from the recorded streams of remotely acquired physical data, these variables serve as inputs to a variety of nonoptical models. Linkages of these models are not necessarily seamless.

In an escalating number of commercially available computer programs, interlocking models may be used to generate maps of variables or produce environmental, financial, health, or cultural predictions that contribute to legislated environmental, economic, industrial, or human control. This further complicates the issue, extends the uncertainties, and often masks the incompatibilities. Many resource managers and research scientists apply these readily available and convenient commercial computer programs to predict and/or assess a priority issue or concern without cognizance of the virtues, limitations, or application restrictions of the models and algorithms selected for inclusion within the program. Detailed discussions on interactive environmental models and development of environmental information and decision-making support systems may be found in the proceedings of the Third International Symposium on Environmental Software Systems ISESS (ISESS, 1999), Lillesand and Kiefer (2000), and elsewhere.

Thus, the role of the ubiquitous models in the plight of aquatic remote sensing is to navigate the highway that leads from remotely sensed water color data to formulated environmental policy. This role may be chronologically recapped: models and algorithms are necessary to

- Remove the color spectrum of the atmosphere from the color spectrum of the water body
- Convert the inferred color of the water body into inferred numerical values of the coexisting concentrations of organic and inorganic CPAs
- Convert these inferred values of CPA concentrations into well-defined status and trends of stress-related response functions
- Distinguish status and trends resulting from the evolution of "naturally" stressed ecosystem behavior from status and trends resulting from "unnaturally" stressed ecosystem behavior
- Utilize inferences emerging from multidisciplinary processes to infer and predict social, cultural, and fiscal consequences — a step invariably outside the sphere of the remote sensing community

The physical, biophysical, and biogeophysical models developed for extracting water quality information from spectro-optical data remotely acquired over optically complex waters are based on sound and robust mathematical formulations (Chapter Two and Chapter Three). However, they are site specific in that they depend on local atmospheric conditions as well as on the optical properties of the indigenous CPAs. These optical properties, in turn, are dictated by local and global geologic diversities. Thus, calibration and validation of the various models become a frustration, particularly when intercomparison of competing models is attempted.

1.9   The inevitable compromise of aquatic remote sensing

It has been historically evident that, throughout the development of physics (the most well-established and mathematically precise of all scientific disciplines), very rarely have theory, model, computer simulation, and observation been in complete agreement. Adding the considerably less than mathematically precise disciplines of biological response and behavior, sociology, finance, and politics to environmental monitoring protocols renders total agreement between any pair of theory, model, computer simulation, and observation a highly elusive dream.

Specifically, the dependence of remote sensing on a variety of interdisciplinary models and algorithms that have as yet to be validated (in some instances, yet to be devised by many outside the remote sensing scientific community) is a major source of uncertainties in remote terrestrial and aquatic environmental monitoring. Scaling issues resulting from local and regional fields of view provide further obstacles and uncertainties. A major source of uncertainties is the dynamic nature of the environmental target. The water column is an ever changing medium — essentially making target replication impossible for successive visits of the satellite or aircraft. Therefore, the calibration and validation of bio-optical and other forward and inverse aquatic models must contend with the realization that precise knowledge of the inferred aquatic variable rarely exists.

Clearly, a compromise must be struck between defendable science and defendable usage of science: how bad can still be good? Understandably, this is a highly subjective issue and universal consensus may forever remain unattainable. Nevertheless, responsibility for this compromise would logically rest within the scientific remote sensing community. Physics and its companion mathematics represent precision, and biochemistry (outside biogenetics and/or rigidly controlled laboratory conditions), human and wildlife behavior, and environmental indices are governed more by statistics than by adherence to physical laws and mathematical formulism. Thus, this compromise between defendable science and defendable usage of science becomes a compromise between physical laws and statistical departures from those laws.

Resolution of these statistical departures comprises the province of the evolving field of environmetrics (El-Shaarawi and Hunter, 2002). In this field, statistical analyses are applied to such topics as sampling schemes and protocols; reliability; statistical accuracy and reproducibility of data sets and environmental models; sensitivity analyses; and general issues of instrument and data calibration and validation. Although environmetrics may provide some necessary tools to assess the compromise, it must be remembered that, like remote and conventional water quality monitoring, environmetrics is also site specific because data collection protocols are site specific and therefore generally issue specific as well.

Remote sensing, however, should not be stringently imposed to provide data that are more reliable than or necessarily equal to the reliability of those acquired by ground-based monitoring systems. Here lies further controversy: namely, the "apples and oranges" comparison between what ground based monitoring can provide and what remote sensing can provide. Without intentionally fueling such controversy, we would maintain that remote sensing is by far more accurate for an entire water body than any feasible sampling scheme. Certainly, for the 1 to 10 L collected *in situ* from a station within the water body, laboratory analyses are more accurate. However, if one considers senescence, bioproductivity, and metabolism as well as storage losses and extraction losses, the pendulum might swing in favor of remote sensing. Understandably, laboratory managers would not agree.

In The Netherlands (A.G. Dekker, private communication), the Water Quality Monitoring Authority (WQMA) imposes higher quality criteria on remote sensing than it does on its own methods. Its objections to remote sensing include

- Inability of remote sensing to comply with Dutch standards of ~150 designated water quality variables (chemical microsubstances) from networks of fixed sampling stations
- Inability of remote sensing to comply with prescribed reporting requirements based on mean values of summer concentrations
- Extra costs associated with remote sensing and its ancillary *in situ* field excursions
- The weather dependence of remote sensing
- Ongoing suspicion of the reliability of remote water quality monitoring

However, in deference to fairness and accuracy, The Netherlands is not alone in its disinterest (and distrust) of inland and coastal water quality monitoring from space. Advances have been made in the science and applications of remotely monitoring case 2 water quality in the U.S., Germany, Canada, Britain, Australia, Switzerland, Japan, Norway, Russia, and elsewhere. Despite this, the reluctance of environmental managers and political policy-makers to incorporate remote sensing into protocols of environmental monitoring and assessment is still very much in evidence (although to varying degrees).

Official sets of standards are absent from remote sensing products — an issue that must be realized and corrected. However, restrictive standards such as those unrealistically imposed by the WQMA of The Netherlands could sound the death knell for remote sensing of inland and coastal water quality. Common ground must be found between the criteria imposed upon vector data (collected *in situ*) and those imposed upon raster data (collected remotely). One global environmental issue is delineation of "ecosystems at risk," a popular (although ill-defined) phrase applied to myriad real or perceived environmental threats. Such predictive issues are very difficult to define in numerically precise values or even in terms of precise risk units other than perhaps percentage or probability. In place of restrictive standards, a much more simple and qualified criterion such as "yes, no, maybe" or "red, amber, green" might be suitable. Early warning is of paramount importance to environmental risk monitoring and assessment. Because real-time or near-real-time remote sensing would therefore be very important to such early warning, it is another consideration of consequence to water quality monitoring from space.

Thus, a second important element of this inevitable compromise is the degree of accuracy, rigor, reproducibility, and/or generality required of the remotely sensed inference of an environmental variable, but also of the directly measured *in situ* value of that environmental variable. Such requirements should logically reflect the needs of the environmental decision-makers and the political policy-makers. However, despite any obvious commonality, these needs are generally articulated on a case-by-case and issue-by-issue basis.

In order to implement this compromise effectively, meaningful dialogue must cohesively interweave technical and scientific developers of the remote sensing reality, providers of the real-time and archival data, the academic community, managers responsible for municipal, national, and international environmental stewardship, and politicians responsible for legislation of such stewardship. Admittedly, such discussions have been undertaken for decades and, clearly, the impasse between the remotely sensed water products and their influential user acceptance has yet to be breached. To do so, we suggest that, as yet, we may not have actually produced water quality "products," but rather some very good science and demonstration of the validity of that science. The compromise requires moving past this demonstration stage to generation of products that are recognized as highly valuable to public interest and concern. Furthermore, to make the science less formidable and suspect to the nonscientific community, a marketing strategy must be devised and successfully implemented. This marketing strategy should revolve around the need to

- Monitor the quality of inland and coastal water quality
- Utilize remote sensing vehicles and sensors as a supplement to, rather than a substitute for, conventional ground-based monitoring networks
- Reach the public media via a focus on the issue rather than a focus on remote sensing
- Somehow overcome the financial inaccessibility of the three decades of archives of time-series satellite data essential for compellingly illustrating the value of remote sensing to recording, understanding, and assessing environmental change
- Make public good a private sector product
- Perhaps, repackage excellent work done by scientists and published in learned journals into applied journals or magazines reaching the targeted end-user community
- Generally, build up a catalogue of value-added remote sensing (VARS) applications

Communication is of paramount importance; however, it comes at a cost and could crush the careers of young scientists because they are not expected to market remote sensing. Their job is to do science. More specifically, however, their job at this stage includes (or should include) applied research into public areas that could benefit from the science that they are producing and also from the science that has already moved them to the stage that they are now advancing.

How does remote sensing attract public attention, media interest, and user (primarily government agencies) awareness of the market value of water quality products? Only two viable options for remote sensing marketing are really available: professional marketers or science liaison personnel at universities or government institutions. A third option might be senior remote sensing scientists who have become frustrated at seeing their work ignored outside the scientific community; however, they are not necessarily versed in advertising and marketing. Professional marketers must first be made aware of the advantages of remotely monitoring water quality and this involves further communication. Science liaison personnel exist at most research-oriented institutes and would therefore appear to be the option of choice. However, the fact that their existence does not appear to have as yet allayed the plight of remote sensing of water quality is disconcerting, emphasizing the low priority given to remote sensing by environmentally focused research institutes.

Remote sensing is a data-gathering tool that observes effect and not cause. Remote sensing scientists are interested in cause, effect, and the science that enables transit from one to the other. The largest groups of "outside" users (scientists not involved in remote sensing, environmental decision-makers, and political policy-makers) are interested solely in the cause and not the effect or the tool. Sufficient overlap between research and product development may or may not be present, although sufficient research does exist to allow dialogue on cause from remote monitoring of water quality. However, "issue-of-the-day" environmental management does have a time line at variance with the time that it takes to attend to it. Research and application budget cuts and staff downsizing resulting from the fiscal challenges of the past decade may have resulted in overtasked and undermanned user agencies. Such a scenario, whether real or perceived, would understandably inhibit the acceptance of untried and unfamiliar additions to water quality monitoring practices — particularly if such additions are discerned as disrupting the continuous data sets (irrespective of *their* value) comprising the time-series archives of the existing monitoring stations.

1.10 Some concluding comments

One of the difficulties inherent to introspection is that it really has no well-defined origin or termination, but rather an unconfined middle that must at some point be reluctantly put on hold. Environmental remote sensing should be believably seen as the gathering and processing of information about the Earth's environment, particularly its natural and cultural resources, from multispectral data obtained from aircraft or satellite altitudes. To date, environmental remote sensing of case 2 water quality has only been *believably* seen by the aquatic remote sensing research community that has performed the necessary science for water quality products that could provide environmental information.

Sad facts emerge from this introspection:

- Of the four categories of end-user groups, the scientific community (remote-sensing and non-remote-sensing scientists) comprises two of these groups.
- The research community also comprises the end-usership of three of the four level products resulting from remotely sensing case 2 waters.
- Level 4 products essential to decision-making and policy formulation require additional models and methods that are invariably beyond participation by the scientific community that has taken level 1 to level 4.

Although these facts might appear to indict the aquatic optics research community as incestuous and self-serving, this is an historical problem that has plagued virtually every scientific discipline and activity that has ventured forward with innovative concepts into areas of long-standing complacency.

Nonetheless, evidence of the relevance of remotely monitoring the quality of inland and coastal water bodies must be expanded beyond the scientific community that is making such remote monitoring possible. This means that decision- and policy-making end-users must buy into this relevance. This has not as yet happened. Despite the inescapable aura of *déjà vu* of this remote sensing retrospection and introspection, we suggest that, in order to make this happen, convincing evidence of this relevance might emerge from essential compromises between defendable remote sensing science and defendable use of that science. Compromises would be determined by the nature and degree of rigor of the environmental information required by the nonscientific end-users of water quality information.

Each new fashion had been hailed as a panacea; "Now we shall vanquish, now the machine will work." Each had gone out without a whimper, leaving behind it the familiar English muddle of which, more and more, in retrospect, he saw himself as a lifelong moderator.

From *Smiley's People*, John Le Carré, 1980

chapter two

Remote sensing of inland water quality: a science primer

Let me see, then, what thereat is, and this mystery explore.

From *The Raven*, Edgar Allan Poe, 1845

This book is directed towards an audience comprising veteran and rookie scientific users of remotely acquired data; undergraduate and graduate students; their scientific mentors; environmental decision-makers; and political policy-makers. Such an audience, albeit one of common goals, is one of diverse backgrounds, interests, and responsibilities. Remote sensing of inland and coastal water quality is integral to their common goals. The science of aquatic optics and radiative transfer mechanisms is integral to the development and validation of the models and methodologies that have generated inland water quality products. In turn, these water quality products have enabled the remote sensing of optically complex natural waters to be considered a valuable component of environmental monitoring.

Water quality products fall within the provinces of the entire intended audience. However, detailed discussions of the governing science often do not. Therefore, in this chapter, remote sensing science will be considered in the form of a brief anecdotal history of how it got us from "there" to "here," i.e., from space science to this point in the saga of inland water quality determination from space. Chapter Three will extend this science "primer" to a "walk-through" of the scientific philosophy and vision that has made remote sensing of complex water bodies an operational reality. As such, it will touch upon the basic principles of photon propagation from the sun and sky through the air–water interface and back to the satellite sensor. Chapter Three is not intended as a short course in aquatic optics. It will, however, discuss the optical properties of the organic and inorganic matter residing within inland and coastal waters and how these optical properties are used to derive mathematical models that relate natural water color to coexisting concentrations of these suspended and dissolved organic and inorganic constituents.

Although each current and potential user of water quality products is invited (and urged) to read Chapter Two and Chapter Three, it is possible that some readers whose interest and background comprise environmental decision-making and policy-formulation may choose to disregard the extended details of Chapter Three. Therefore, moving directly from Chapter Two to Chapter Four would not result in loss of generality or continuity.

2.1   Remotely sensing the Earth from space

The mid-1950s witnessed a rapidly escalating scientific and public interest in space exploration that spanned the spectrum from Buck Rogers-like galaxy-travel whimsy to the much more serious and focused race to the lunar surface. The 1950s and 1960s were exhilarating decades for science and technology. The ongoing development of space hardware (rockets, sophisticated sensor packages, high-speed computers and telemetry systems) and formulation of mathematical models of interplanetary dynamical processes benefited not only the unmanned missions of the space program, but also the manned missions. The concept of "space weather" — the intensity, propagation, and directionality of biologically pernicious solar radiation within the inner solar system — provided a very logical linkage between the two missions; one result of this was delayed Apollo launchings on a number of occasions due to enhanced fluxes being recorded by cosmic ray sensors at various locations within the inner solar cavity.

The intellect and creativity of a dedicated and enterprising mixture of established and young scientists used the wealth of telemetered data from a cluster of Earth-orbiting satellites and deep-space probes effectively to study and monitor heliophysical and electromagnetic configurations within the inner solar system, for example:

- Solar flare impacts on terrestrial magnetic fields
- van Allen radiation belts around the Earth
- The solar "wind"
- The 11-year solar activity cycle
- Co-rotating Forbush decreases
- Solar and galactic cosmic ray propagation and modulation

- Energetic solar storm particle events
- Filamentary microstructures within solar magnetic fields

Many more examples exist. Space science was truly a heady, vibrant, and exciting field!

However, turbulent demonstrative activism resulting from major re-evaluations of social, political, cultural, and moral behavior also dominated the 1960s agendas of developed and some developing nations. One aspect of this turbulent activism was the "future shock" vision of inevitable destruction (by human hand) of the Earth's environment. Perhaps as an effort to make space science more relevant to such socially troubled times, NASA (soon to be followed by the space agencies of other countries) decided that the planets that it was planning to study must include Earth. Thus, environmental monitoring from space was born.

Unquestionably, environmental monitoring from space was/is a noble pursuit. Such pursuit, however, presented some very stubborn challenges to space and environmental science. These challenges arose from differences between the respective scenarios of space science and environmental monitoring. Space science has the major advantage of being a pursuit essentially focused and dependent upon a single discipline, namely, physics. The planets, suns, galaxies, interplanetary and intergalactic media — indeed, the entire universe — comprise an awesome dynamic equilibrium governed by rigid adherence to precise, mathematically defined laws of gravity and spectroelectromagnetism. Satellite payloads of Earth-orbiting and deep-space probe missions included sensors designed to:

- Directly measure strength and direction of magnetic fields
- Identify atomic and nucleonic particle types and determine their velocities, flux densities, and directions of motion
- Measure plasma densities and velocities directly
- Measure directly a variety of other space parameters that could be fused into models to describe the interactions of solar and galactic radiation with the Earth's magnetic field, planetary atmospheres, and the interplanetary medium

Thus, within the scenario of space science, direct measurements of *physical* parameters were used to derive models that defined the behavior of a system of particles constrained by *physical* laws. Ability to use physics to infer more physics provides a powerful and precise scientific tool.

By contrast, environmental monitoring of Earth and any other Earth-like planets that might exist is, of necessity, a multidisciplinary activity. Physics, chemistry, biology, and behavioral sciences interact in complex fashions defying readily recognizable adherence to mathematical formulism. Here too, however, satellite sensors measure *physical* parameters (reflected and/or self-emissive spectral radiation from atmospheric, terrestrial, aquatic, and wetland features comprising Earth's biosphere). As has been, and will continue to be, emphasized throughout this book, environmental decision-makers and political policy-makers require information on the *biological* status of these biospheric components. This need to use physics to infer biology results in the irony of environmental remote sensing that was introduced in Section 1.4.

Therefore, bio-optical modeling has been the driving force behind the environmental remote sensing research pursued by university, government, and private sector scientists since the late 1960s. The rest of this chapter will narrate the development of such models by the aquatic optics and remote sensing research community. Specific focus will be on the scientific philosophy that enables water quality information to be extracted from remotely acquired inland and coastal water color data.

2.2 Remotely sensing aquatic color from space

Environmental satellites contain a variety of sensors that are sensitive to various regions of the electromagnetic spectrum. Although other regions of this spectrum (most notably microwave, radar, and thermal infrared) have been and are being represented in satellite and airborne packages dedicated to aquatic monitoring, spectral measurement of water color is essential to remotely monitoring water quality. Herein, therefore, we will restrict our focus to the visible light region of the electromagnetic spectrum (i.e., radiation contained within the boundaries of the color wavelengths, namely, from ~390 to ~740 nm, from blue to red). Furthermore, the principal sensors used to monitor aquatic color (imaging spectroradiometers) are passive devices, i.e., devices that faithfully respond to the optical field emanating from a target irrespective of whether that optical field is "reflected" (i.e., from light originating at the sun) or self-generated (i.e., from light or other radiation originating at the target). For simplicity, we will consider only that fraction of the visible solar and sky radiation recorded at the remote platform as a consequence of its being "reflected." Radiation that is self-generated by the water will not be considered here.

The color of a natural water body is governed by the spectral dependencies of in-water absorption and scattering of downwelling global radiation (direct solar and diffuse skylight) by the composite of indigenous organic and inorganic color-producing agents (CPAs) coexisting within the water column at the time the remote observation is performed. In addition to molecular "pure" water, these aquatic CPAs are generally considered to be phytoplankton, suspended sediments, and dissolved organic matter. The coextant concentrations of these CPAs are taken as the remote sensing definition of water quality.

The terms *phytoplankton, suspended sediments,* and *dissolved organic matter* do not each represent a single species of CPA, but rather are collective terms representing the genre of principal participants in the absorption and scattering of incident radiation. Each genre contains a variety of distinct species that, while maintaining individual properties, also displays the general properties of its genre. Chlorophyll *a* is generally considered a quantifiable surrogate for phytoplankton, suspended minerals a quantifiable surrogate for suspended sediments, and dissolved organic carbon a quantifiable surrogate for dissolved organic matter. Of prime importance to the generation of remote sensing water quality products is the estimation from space of chlorophyll concentration.

2.3 Remotely sensing inland and coastal water quality from space

Unambiguously converting remote determinations of water color into water quality information is far from a trivial scientific task. For the optical complexity generally characteristic of inland and coastal (the so-called case 2) waters, the task can be and has been quite formidable.

Some obvious questions facing the scientific community self-charged with converting case 2 water *color* into case 2 water *quality* included:

- What water quality parameters could be extracted from water color?
- What optical processes are responsible for aquatic color and how may these processes be expressed mathematically?
- How are the so-called case 2 (most inland and coastal) waters optically different from the so-called case 1 (mid-ocean and some inland) waters?
- How does a lake or coastal bottom substrate visible from a remote platform affect the color spectrum recorded at the remote platform?
- How does a vegetative canopy on such a visibly apparent substrate affect the color spectrum recorded at the remote platform?
- How does the intervening atmosphere obfuscate the "true" color of the water being remotely monitored? How can this obfuscation be removed?
- How are the atmospheres above inland and coastal waters similar to and different from one another? How are they similar to and different from those above mid-ocean waters?
- What is known and what more needs to be known about the optical properties of organic and inorganic matter resident within inland and coastal waters?
- How do geography and seasons exert an impact on indigenous aquatic optical properties?
- How are relationships among the organic and inorganic constituents (CPAs) and aquatic color expressed mathematically?

The scientific quest to answer these questions (and others as they emerged during the quest) consisted of logical and cohesive interplay between technology and science: namely, the development and deployment of satellite, aircraft, and *in situ* optical sensor systems and the formulation and application of spectro-optical and spectral bio-optical mathematical models.

In addition to molecular water, the principal CPAs for the relatively clear mid-ocean waters (for which optical modeling has been a research topic throughout the past century) are the resident chlorophyll-bearing biota (phytoplankton) and their co-varying degradation products (detritus). Thus, water quality modeling of mid-ocean water quality conveniently reduces to a single-parameter problem; that parameter is chlorophyll, which, fortuitously, is the water quality indicator most commonly sought by environmental decision-makers and of prime importance to policy-makers. However, in addition to phytoplankton, detritus, and molecular water, principal CPAs for the optically complex inland and coastal waters include a wide variety and concentration range of suspended and dissolved organic and inorganic matter of terrestrial origin — CPAs that seldom, if ever, co-vary. Thus, water quality modeling of inland and coastal waters becomes a non-linear, multiparameter problem.

Therefore, in order to extract the sought after concentration of chlorophyll from a remote spectral measurement of inland or coastal water color, it also becomes incumbent to extract the concentrations of its co-CPAs simultaneously. In some instances, chlorophyll may indeed be the principal CPA. In other instances, dissolved organic matter may be the principal CPA. In a majority of instances, however, the suspended sediments are the overwhelmingly dominant CPAs, indicating that, as suggested by R.H. Stavn (private communication), bio-optical models should be more appropriately considered as bio–geo-optical models.

The radiation recorded by a remote sensing device is a composite of photons of differing histories. Figure 2.1 illustrates the familiar pathway partitioning of the visible spectral radiation (light) that originates at the Sun, propagates to the Earth, downward through the atmosphere, to and through the air–water interface, within the water column, and then upward — ultimately reaching and being recorded by the remote sensing device. The radiation that eventually is recorded at the satellite orbit may contain as many as four distinct components. These are simply labeled in the figure as:

- $[R_A(\lambda)]$: that portion of the recorded spectral radiation, $R(\lambda)$, resulting from solar and sky radiation that never reaches the air–water interface. It is light that has been single-scattered or multiple-scattered by the atmosphere and represents the color of the atmosphere at the time of satellite fly-by.
- $[R_{AW}(\lambda)]$: that portion of the recorded spectral radiation, $R(\lambda)$, resulting from solar and sky radiation that reaches but does not penetrate the air–water interface. It is light that has been reflected from the air–water interface and represents the color of the air–water interface at the time of satellite fly-by.
- $[R_W(\lambda)]$: that portion of the recorded spectral radiation, $R(\lambda)$, resulting from solar and sky radiation that penetrates the air–water interface,

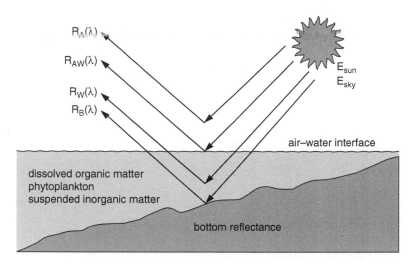

Figure 2.1 Components of the radiance recorded by a remote sensor over a natural water body.

interacts with the water volume, and re-emerges without encountering the bottom of the water column. It therefore represents the color of the water column at the time of satellite fly-by.

• $[R_B(\lambda)]$: that portion of the recorded spectral radiation, $R(\lambda)$, resulting from solar and sky radiation that penetrates the air–water interface, reaches the bottom of the water column, and re-emerges from the water column. It is light that has been reflected from the bottom surface and represents the color of the bottom surface.

Conceptually, therefore, the visible spectral radiation recorded at the satellite (the color of the biosphere immediately below), $R(\lambda)$, would be given as the sum of $R_A(\lambda)$, $R_{AW}(\lambda)$, $R_W(\lambda)$, and $R_B(\lambda)$. Then the color of the water body would be obtained from:

$$R_W(\lambda) = R(\lambda) - [R_A(\lambda) + R_{AW}(\lambda) + R_B(\lambda)] \qquad (2.1)$$

Equation 2.1 considers the returns from the atmosphere, air–water interface, water column, and bottom to be independent entities. Clearly, however, they are not. The return from the atmosphere is a component of each of the other three returns, and the return from the air–water interface is a component of each of the two water-penetrating returns. This reality notwithstanding, mathematical models have been established to separate atmospheric, surficial, water volumetric, and bottom returns from the radiation spectrum recorded by the satellite. This is not surprising because, to this point, physics has been employed to infer more physics.

Scattering and absorption of light photons by atmospheric and aquatic matter readily lends itself to precise mathematical formulism. Such is certainly the case for atmospheric optics and aquatic optics. Both scientific topics are fundamental to aquatic remote sensing. Remote sensing of water is essentially the viewing of one attenuating medium (water) with one set of specific optical properties through a second attenuating medium (atmosphere) with a second set of specific optical properties.

2.4   The scientific methodology (forward and inverse models)

In Chapter One, we emphasized the dependence of remote sensing in general and environmental remote sensing in particular on models to generate a hierarchy of remote sensing products of ever increasing complexity. Conceptual thinking and mathematical prowess are two of the premium attributes required for successful modeling. As propounded in Chapter One and again herein, physics is a precision discipline in which precise mathematical formulism becomes a routine feature of physical models (i.e., models that utilize physics to infer more physics). Models dealing with atmospheric optics and atmospheric photon propagation, as well as models dealing with aquatic optics and aquatic photon propagation, are physical models.

The physics base of such model development is the theory of transfer of energy. The mathematical physics base of such model development is Monte Carlo simulations of photon propagation within attenuating media. A Monte Carlo simulation is a semiempirical means of linking the optical properties of an attenuating medium to the properties of the radiation field within that medium. Such an approach is ideally suited to the stochastic nature of individual photon propagation. A high-speed computer is used to follow the propagation of a very large number of photons, considered individually, through an imaginary volume of water (or gaseous medium) possessing a preselected set of optical properties. Random numbers are generated to select, in accordance with these preselected properties, a variety of conditions to affect the fate that the photon encounters when interacting with the absorption and scattering centers within the attenuating medium or media. Such Monte Carlo computer programs have been effective in establishing an impressive science base for aquatic optics and remote sensing. A partial list of emergent modeling successes would include:

- Mathematically relating the apparent optical properties of a water column (i.e., optical properties that depend upon the spatial distribution of the impinging radiation) and the inherent optical properties of a water column (i.e., optical properties that are totally independent of the spatial distribution of the impinging radiation)
- Mathematically relating the bulk inherent optical properties of a water column (i.e., the inherent optical properties when the water column is considered as a composite entity with no regard as to the specific components contributing to that property) to the specific

inherent optical properties of a water column (i.e., the inherent
optical properties that may be attributed to the individual organic
and inorganic absorption and scattering centers comprising a water
column)

- Mathematically describing the transference of radiation between the
water body and the atmosphere
- Mathematically relating the inherent optical properties of the water
column and the water-leaving radiation (i.e., upwelling radiation
emergent from the water column subsequent to its interactions with
subsurface aquatic matter; this water-leaving spectral radiation defines the color of the water body) to the coexisting concentrations of
the indigenous organic and inorganic matter

Specific inherent optical properties (IOPs) of the organic and inorganic
matter resident within the water column provide the linkages between
remotely sensed water color and the coexisting concentrations of the matter
responsible for that water color. The values of these concentrations (generally
taken to be those of chlorophyll, suspended sediments, and dissolved organic
matter) comprise the water quality indicators to be extracted from remotely
sensed water data. Specific IOPs (often referred to in the limnological literature as *optical cross section spectra*) are defined as the specific amount of
absorption and scattering, as a function of wavelength, that may be attributed to a unit concentration of each organic and inorganic component of a
natural water body, including the molecular water. The upwelling radiation
(color) emergent from the water has been mathematically expressed as a
function of the specific IOPs and the concentrations of the coexisting
color-producing agents, CPAs (namely, molecular water, chlorophyll, suspended sediments, dissolved organic matter). Thus, if the IOPs and color of
the water are known or can be judiciously inferred, the concentrations of the
CPAs may be reliably determined. Similarly, the specific IOPs may be reliably
determined from precise values of the CPA concentrations and the water
color.

Determining the specific IOPs (optical cross section spectra) for a natural
water body is generally accomplished in one of two ways:

- In laboratory determinations, a species and concentration of organic
or inorganic CPA are preselected. The nature of the containment
medium is controlled within precise laboratory protocols. Optical
devices are then used to measure precisely the absorption and scattering properties inherent to the selected CPA.
- From field surveys, *in situ* measurements of the subsurface radiation
field are performed in concert with the collection of water samples
that are returned to the laboratory for accurate determinations of the
concentrations of chlorophyll, suspended sediments, and dissolved
organic matter (usually carbon).

Not surprisingly, these two methods do not yield identical results. This will be discussed further in Chapter Three. Clearly, however, because the inherent optical properties of indigenous aquatic colorants are functions of geologic diversity and of time (see following), obtaining the specific IOPs appropriate for inland and coastal water bodies on a global scale becomes site specific and, therefore, a labor-, time-, and cost-intensive task. Thus, an unfortunate lacuna that must be filled still remains.

Modeling water quality and water color involves *forward* modeling and *inverse* modeling. Forward models utilize IOPs and CPA concentrations to generate water color. Inverse models utilize IOPs to deconvolve water color into the CPA concentrations responsible for that color. The principal goal of forward and inverse water color models is to obtain information on the quality of the water being remotely monitored, i.e., to obtain, simultaneously, concentration values of the indigenous chlorophyll (to evaluate changes in the biological status of the water), suspended sediments (to evaluate changes in aquatic turbidity), and dissolved organic matter (to evaluate the roles of inland and coastal waters — for example, in carbon cycle dynamics).

Reiterating, the Monte Carlo simulations, forward models, and inverse models are physical models inter-relating, in mathematical equation form, variables from the physics fields of optics, radiometric color, electrodynamics, and energy transfer. Thus, these inter-relationships possess high degrees of precision. As precise as a mathematical equation may be, however, accuracy of its numerical output is dictated by accuracies of the inputted numerical values of the variables that it inter-relates (the classic "garbage-in–garbage-out" syndrome). Therein lies the rub.

The Earth's environment is in a perpetual state of flux, a complex phenomenon of stationary, quasistationary, periodic, nonperiodic, random, predictable, and/or unpredictable change. Also, such change varies in geographical location and time. Numerical values of the physical parameters of optical radiation, CPA concentrations, and optical cross section spectra may be accurately determined for a specific site area at a specific time. However, the indigenous species of chlorophyll-bearing biota and, therefore, the optical cross section spectra of chlorophyll are consequences of the trophic conditions of the water body. The indigenous species of aquatic matter of terrestrial origin (e.g., suspended sediments and dissolved organic carbon) are consequences of the geological and climatic aspects of the basin containing the water body. Obtaining pertinent optical cross section spectra are field- and laboratory-intensive activities. Thus, precise spectral values (including their temporal variability) have only been determined for a limited number of site-specific studies.

The catalogue of inland water inherent spectro-optical properties must certainly be expanded. However, recent research has suggested that common (or at least similar) optical cross section spectra could be utilized as surrogate values for watersheds displaying comparable geologic origin and/or history. In some instances, therefore, such surrogate values *might* be judiciously used in optical and bio–geo-optical models to produce water quality products that were environmentally consequential and scientifically detendable.

Unfortunately, vagaries of atmospheric CPA concentration profiles pose much greater problems to water quality determinations from space than do vagaries of indigenous aquatic optical cross section spectra. Atmospheric attenuation results from Rayleigh scattering (atmospheric molecules), Mie scattering (aerosols), and absorption (water vapor, ozone, gases such as oxygen, nitrogen, and carbon dioxide). Mathematical equations defining Rayleigh scattering, Mie scattering, and absorption have been derived based upon sound science. Numerical values of absorption and Rayleigh scattering cross section spectra may be accurately represented in models. However, the Mie scattering component of atmospheric attenuation is still an unresolved problem. Atmospheric aerosols are suspended atmospheric solid and liquid particles that exist in a multitude of diverse forms and shapes (smoke, water, clouds, ashes, dust, pollen, spores, and acids); the perverse and intractable patchiness of these aerosols has stymied precise determination of their distribution profile between the satellite and the targeted water body.

Frustrations besieging attempts to extract water quality information by removing atmospheric interference from satellite water color data are illustrated in Figure 2.2 and Figure 2.3. Both figures display August 1, 1999 SeaWiFS imagery of the Gulf of Finland (provided courtesy of Dimitry V. Pozdnyakov, Nansen International Environmental and Remote Sensing Center). Land radiances have been excised from the imagery. St. Petersburg, Russia, is located at the farthest eastern extreme of the Gulf waters. Helsinki, Finland, is located about midway along the northern shore. The water body to the south of the Gulf is Lake Chudskoya, Estonia.

The SeaWiFS level 1 data were processed into level 2 water quality products using a common bio–geo-optical model with two different standard atmospheric correction techniques. Figure 2.2 compares chlorophyll maps emergent from the use of each of the atmospheric correction algorithms. Figure 2.3 compares the chlorophyll map arising from the use of one atmospheric algorithm with the suspended sediment map emerging from the use of the other. Two very disconcerting features become immediately obvious:

- The highest chlorophyll concentration patterns in the Gulf of Finland emerging from the use of one atmospheric correction algorithm are delineated as the lowest, essentially absent chlorophyll concentrations arising from the use of the other (Figure 2.2).
- The chlorophyll concentration map that emerged from the use of one atmospheric correction algorithm is essentially identical to the suspended sediment map arising from the use of the other (Figure 2.3).

Both atmospheric algorithms depend upon accurate information on atmospheric aerosol concentration profiles immediately above the targeted water body. Figure 2.2 and Figure 2.3 illustrate that neither may have had that information. This inability to obtain precise numerical values of local

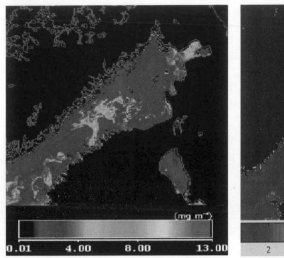 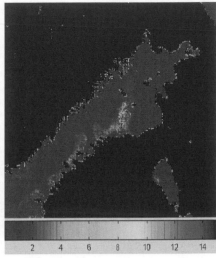

| (a) Atmospheric algorithm one | (b) Atmospheric algorithm two |

Figure 2.2 (See color insert following page 134.) Aquatic chlorophyll maps acquired from August 1, 1999, SeaWiFS data over the Gulf of Finland using the same aquatic bio–geo-optical model with two different atmospheric correction algorithms. (Mapping provided courtesy of Dimitry V. Pozdnyakov, Nansen International Environmental and Remote Sensing Center.)

above-water aerosol concentrations results in the most difficult aspect of water quality monitoring from space being the removal of unwanted atmospheric signal from the environmental color recorded at satellite altitudes. Atmospheric intervention will be discussed further in Chapter Three.

2.5   Concluding remarks

Remote sensing of aquatic color was initially developed for mid-ocean waters, the varying optical properties of which are determined solely by phytoplankton and the co-varying products of their mortality. The optical simplicity of these case 1 waters has enabled ocean chlorophyll concentrations to be inferred remotely with accuracies deemed acceptable for most oceanographic biological purposes. Furthermore, such accuracy has been obtained using semiempirical, unsophisticated algorithms involving simple ratios of two or three wavelength bands in the visible spectrum. In most instances, a full color spectrum, although certainly desirable, is not essential for remotely extracting water quality information (namely, chlorophyll concentrations) from case 1 waters.

The cumulative surface area of continental waters is essentially negligible when compared to the surface area of intercontinental waters. Cumulative volumetric intercomparison becomes even more dramatic. However, as will be discussed in Chapter Four, most of the world population resides

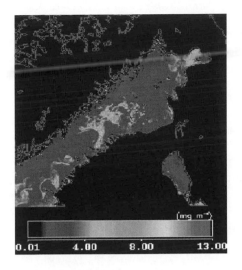 

(a) Surface chlorophyll map
atmospheric algorithm one

(b) Suspended sediment map
atmospheric algorithm two

Figure 2.3 (See color insert following page 134.) A comparison of the chlorophyll map emergent from the use of one atmospheric algorithm with the suspended sediment map arising from the use of a different atmospheric correction algorithm on the August 1, 1999, SeaWiFS image over the Gulf of Finland. The same bio–geo-optical multivariate model was used subsequent to Level 1 radiance correction by each atmospheric algorithm. (Mapping provided courtesy of Dimitry V. Pozdnyakov, Nansen International Environmental and Remote Sensing Center.)

along or close to the shores of continental or intercontinental waters. It is these inland and coastal waters that provide the water essential to sustain the drinking, cleaning, industrial, power, and recreational needs of the Earth's human and wildlife populations. These inland and coastal waters are also most vulnerable to the toxic and nontoxic discharges and disposals that characterize an industrialized society. The world population and its demands on the water supply are increasing. Such escalating use and abuse of a limited natural resource place sustainable health at great risk. The quality of inland and coastal waters and their responses to environmental stress become increasingly important monitoring concerns.

The past decade has seen a sharp emphasis on Earth-orbiting satellites as vehicles for monitoring the quality of inland and coastal (primarily case 2) waters, the optical complexity of which is attributable to adjacent or encapsulating land masses. Obviously, a water system with a variety and wide concentration range of organic and inorganic aquatic colorants that display temporal and spatial variations is more optically complex than mid-ocean water systems. Case 2 waters therefore require models and algorithms of higher sophistication than those extracting chlorophyll concentrations from remote measurements of ocean color. (Apart from the molecular water, chlorophyll is the single colorant in the latter.)

It would intuitively appear that case 1 waters represent the simple limit of case 2 waters (i.e., waters in which the multivariate organic and inorganic suspended and dissolved CPAs reduce to chlorophyll as the determinant CPA). However:

- Ocean optics and ocean bio-optics are scientific disciplines that date at least as far back as the late 19th century.
- The initial foci of environmental remote sensing from satellite plat-forms were terrestrial and ocean targets, with large lakes and coastal zones, at best, a postscript.
- A wealth of research in optical oceanography and atmospheric phys-ics had been and was being conducted in many countries prior and subsequent to the launch of ERTS-1.
- Lake optics and remote sensing of inland and coastal waters was barely in its infancy in the 1970s.

Thus, we were well aware of the role that science played in ocean optics. Sadly, however, we were not equally aware of the role that science played in lake optics.

This historic and somewhat enigmatic lack of focus on lake optics gave rise to a well-meaning, although totally misguided, popular philosophy that the case 1 models and algorithms serving the ocean community so well could, by merely implementing appropriate "tweaks," convert these case 1 water models and algorithms into case 2 water models and algorithms. Although such misconceptions are rapidly disappearing, they have not, as yet, been totally discarded. The forward and inverse optical and bio-optical/bio–geo-optical modeling of inland and coastal waters to which we have very briefly alluded here are based upon sound science. The mathematical relationships are robust and accurate. Case 1 waters are indeed the simple limits of case 2 waters, and the case 2 water models and algorithms contain the case 1 water models and algorithms when a single colorant (chlorophyll) is considered. Of course, we steadfastly advocate that the converse is invari-ably fraudulent, and that case 2 waters should never be considered the "tweaking" limits of case 1 waters. After such an affirmation, however, the following two paragraphs risk being interpreted as scientific betrayal at worst or ignorance at best. Needless to say, we intend neither.

In Chapter One, we have strongly suggested that a compromise was essential between defendable science and defendable usage of that science. In Chapter Four, we will discuss some needs of environmental deci-sion-makers and political policy-makers, and the possible contribution remote sensing of inland and coastal waters might make to them. The science that has related water color to water quality is sound. The optical properties and bio–geo-optical relationships of inland and coastal waters are adequately understood and interpretable. Environmental needs of decision-makers and policy-makers are also sound and interpretable. Many of these needs require application of the full range of case 2 water science for remote sensing to be of consequence to them. The required sophistication of the science is directly (but not linearly) related to the degree of required accuracy as well as the degree of aquatic complexity.

However, for many user needs, the full range of case 2 water science, although welcome, is not essential. Relative values or pattern changes or appearance/disappearance of aquatic features as recorded remotely can often be invaluable to environmental stewardship. For example, optical simplicity of mid-ocean waters has allowed simple ratio algorithms to monitor ocean chlorophyll over extensive surface areas. Although such case 1 algorithms are inappropriate for most case 2 waters, some water quality user needs (1) concern inland and coastal waters that are very similar to case 1 waters; (2) do not necessarily require the robustness of case 2 water science; or (3) both. In such instances, use of the techniques and algorithms that have served the ocean remote sensing community so well may have merit. Perhaps, therefore, "controlled tweaking" in a science/usage of science compromise may have an important role. It cannot be overemphasized, however, that whatever the possible concessions made to this compromise, the integrity of the science must remain uncompromised.

Numerous water-focused sensors (color and beyond color) have demonstrated their value for aircraft and satellite monitoring of inland and coastal water quality. Nonetheless, more sophisticated sensors are being launched and more are being planned. The logical progression is continuing, at least into the near future: namely, that technology leads science and science leads application of science. There is no doubt that the gap between technology and application is ever-increasing. Such a gap, if allowed to expand unfettered, would eventually prove fatal, not only to remote sensing of inland and coastal waters, but also to environmental remote sensing in general.

One cannot rationally argue against the advance of technology, nor should one. Presumably, however, one should be able to argue rationally against the rapidly expanding gap that exists between technology and its scientifically sound application. Science faced the challenge that extracting water quality information from remote measurements of case 2 water color presented to it, and, in a surprisingly short number of years, managed to develop precise and readily defendable optical and bio–geo-optical models and algorithms. The same can be said for optical models and algorithms that describe the energy transfer phenomena responsible for photon propagation within the atmosphere. Unfortunately, however, attempts to use such energy transfer models to remove unwanted atmospheric signals from the environmental color spectrum recorded at the satellite orbit are invariably impeded by the lack of detailed aerosol profiles within the intervening atmosphere. Thus, obtaining site- and time-specific knowledge of these atmospheric aerosol profiles remains an essential concern of remote sensing missions.

The aquatic models and algorithms generated by a dedicated community of theoretical and applied limnological and remote sensing scientists have enabled the visible radiation spectrum emergent from optically complex multivariate water bodies to be converted into water quality products (the coexisting concentrations of indigenous organic and inorganic matter responsible for water color) of consequence to environmental monitoring and decision-making. If these water quality products do not find an influential end-usership, the "gap" wins and remote sensing loses. Inarguably, searches for the "ultimate space vehicle," the "ultimate sensor," the "ultimate optical model," and the "ultimate algorithm" are noble goals worthy of pursuit. Technology will always lead science and science will always lead application of science — this quite simply defines the "nature of things."

However, while these noble goals have been and are being pursued, a large reservoir of some pretty good space vehicles, sensors, optical models, and algorithms to go with some pretty good archives of environmental space data have been inexplicably left with an untapped potential for monitoring and assessing the temporally and spatially changing features of global inland and coastal waters. Tragically, pursuit of ever more sophisticated technology is regarded as more essential, more fun, and more financially rewarding than pursuit of environmental applications. This is a situation that must now be addressed. We have the technology. We have the science. We have the water quality products. We have an urgency to resurrect the purpose for all this.

Thus, at this juncture, it is necessary to

- Harness the scientific vision, excellence, integrity, and doggedness that have successfully attacked and tamed the complex challenge to extract inland water quality from inland water color
- Unleash them on the current remote sensing challenge — namely, that of convincing end-users that this historic science odyssey has not been in vain

A fork looms in the road before us — a path to quietus and a path to life support. The choice would appear to be obvious.

> We'll plug along on it for two or three years, and maybe we'll get something permanent — and probably we'll fail.

> **From *Arrowsmith*, Sinclair Lewis, 1932**

> It has become my observation that people often work on something for years and then some urgent situation comes up…and it suddenly comes through.

> **From *In Search of Excellence*, Thomas J. Peters and Robert H. Waterman, Jr., 1982**

chapter three

The science of remotely sensing case 2 water quality

I often say that when you can measure what you are speaking about, and can express it in numbers, you know something about it; but when you cannot express it in numbers, your knowledge is of a meager and unsatisfactory kind; it may be the beginning of knowledge, but you have scarcely, in your thoughts, advanced to the stage of Science, whatever the matter may be.

William Thomson (Lord Kelvin), circa 1865

The greatest discrepancies in the spectral inherent optical properties (IOPs) of case 2 waters are attributable to geologic diversities. The "forgotten man" in the saga of case 2 water science is the suspended inorganic particulate concentration of local origin. This is a consequence, perhaps, of a hitherto all too prevalent philosophy, that by simply "tweaking" analyses applicable to case 1 waters (in which terrestrial particulates are conveniently absent), those analyses will be rendered applicable to case 2 waters.

One aspect of such "tweaking" was to lump everything that was not chlorophyll into an ill-defined "detrital" term, ascribe colored dissolved organic matter (CDOM) absorbing properties to this term, and correct the total attenuation coefficient value for backscattering. Consequently, there currently exists but an extremely limited number of directly-measured spectral IOPs of suspended inorganic matter in inland and coastal water bodies and the limited number that do exist are far from identical in spectral intensity and spectral shape (absorption and scattering). It is certainly time that this lacuna be addressed. Suspended inorganic matter is an independent color-producing agent (CPA) that very often is the dominant CPA of lakes and rivers and must be thus considered.

Therefore, in order to extract the concentration of a desired aquatic component (most often chlorophyll) from a remote determination of inland or coastal water color, the coexisting concentrations of its coextant optically intensive CPAs (DOC, SOM, and suspended minerals) must be simultaneously extracted. To do so requires knowledge of the spectral inherent optical properties of all aquatic color-producing agents. Case 1 waters may be logically considered to represent a simple limit of case 2 waters; however, case 2 waters should never be dismissed as merely a tweaking limit of case 1 waters.

3.1   Aquatic optics and water color

Aquatic optics is that branch of physics that deals with the interaction of spectral radiation with and its propagation within molecular water and whatever matter is suspended or dissolved therein. Aquatic color is defined by the restricted range of visible radiation (referred to as *light*) within the wavelength interval 390 to 740 nm. The color of natural water bodies is a complex optical feature resulting from in-water scattering and absorption processes, as well as emission by the water column and reflectance from the bottom substrate if that substrate is visible from a vantage point above the air–water interface. In addition to the water, in-water CPAs comprise the resident suspended or dissolved matter and may be organic or inorganic. Variations in water color result from changes in these indigenous CPAs, changes in the concentrations of these indigenous CPAs, and/or changes in both CPAs and their concentrations.

Aquatic optics has evolved along two avenues according to whether the water body was saline (oceanographic) or fresh (limnological). Optical oceanography has benefited from centuries of increasingly sophisticated theoretical and *in situ* research directed towards understanding the radiative transfer mechanisms that govern the establishment of the light regime beneath the air–water interface of oceans. Many excellent treatises and textbooks chronicling advances in optical oceanography exist and form essential reading for researchers and students of aquatic optics. Such works would include the monumental five-volume treatise on hydrological optics by Preisendorfer (1976a–e) and the classic books by Jerlov (1976), Kirk (1983, 1994a), Maul (1985), and others.

Somewhat enigmatically, however, limnological optics is in relative infancy despite the fact that the radiative transfer theory forming the basis of optical oceanography also is relevant to optical limnology. With the advent of environmental satellites, sharp focus was placed on remote sensing of oceanic water quality. Enigmatically, however, very little focus was directed towards remote sensing of limnological water quality.

For saline or fresh waters, however, two principal terms of consequence to understanding photon propagation entering, interacting with, and exiting from the water body are *radiance* and *irradiance*. Although these terms refer to the number of photons at a given site within the aquatic medium, each represents a different subset of the totality of photons at that site. For a spherical geometry of polar angle θ and zenith angle Φ, the radiance at wavelength λ, $L(\theta,\Phi,\lambda)$, is defined as the radiant flux per unit solid angle $d\Omega$ (the solid angle lying along a specified direction) per unit area dA (the area lying at right angles to the specified direction) at any point in the radiative field. The irradiance, $E(\lambda)$, is defined as the radiant flux per unit area at a point within a radiative field or at a point on an extended surface.

The directional aspects attributable to radiance are not attributable to irradiance because irradiance is composed of all the radiant flux impinging at a selected point within the radiative field. Nonetheless, a quasidirectionality is assigned to irradiance because upwelling, $E_u(\lambda)$, and downwelling, $E_d(\lambda)$, irradiances are conveniently considered to be distinguishable entities; each entity defines the totality of impinging photons within a hemisphere of the geometric spheroid centered by the selected point within the radiative field. A third important aquatic optics term, *irradiance reflectance* at a wavelength λ, $R(\lambda)$, is then defined as the ratio of this upwelling to this downwelling irradiance. For the special case of underwater reflectance at depth z, this reflectance is termed the *subsurface reflectance* or the *volume reflectance*, $R(z,\lambda)$, i.e.,

$$R(z,\lambda) = E_u(z,\lambda)/E_d(z,\lambda). \tag{3.1}$$

The value of the subsurface volume reflectance just beneath the air–water interface, $R(0^-,\lambda)$, may be converted into the *water-leaving radiance*, $L_u(0^+,\lambda)$, through the relationship given by Austin (1974):

$$L_u(0^+,\lambda) = R(0^-,\lambda)[1 - \rho(\theta)]E_d(0^+,\lambda)(1 - \rho_{irr})/Qn^2[1- 0.48R(0^-,\lambda)] \tag{3.2}$$

where

$\rho(\theta)$ = internal reflectivity for the in-water refracted angle θ corresponding to the remote sensing viewing angle

ρ_{irr} = surface reflectivity for downwelling irradiance in air

n = relative index of refraction of water to air

$E_d(0^+,\lambda)$ = downwelling irradiance just above the air–water interface

Q = ratio of the upwelling irradiance just beneath the air–water interface to the upwelling nadir radiance just beneath the air–water interface, i.e., Q is defined as $E_u(0^-)/L_u(0^-)$. For most inland and coastal waters, Q can vary in value from ~2.4 to ~5.6 as the solar zenith angle increases from 0 to 80°

(Bukata et al., 1988a)

The radiometric color of a water body as viewed above the air–water interface is then defined as the spectral values of the water-leaving radiance (or
volume reflectance) throughout the wavelength interval $390 \, nm \leq \lambda \leq 740 \, nm$.

Volume reflectance (radiometric color) is an apparent optical property
(AOP) of a natural water body. Apparent optical properties depend upon
the angular and spatial distribution of the impinging radiation as well as
upon the nature and concentrations of matter resident within the water.
Another common apparent optical property of a natural water column is the
diffuse attenuation coefficient for downwelling irradiance, $K_d(z,\lambda)$, which
defines the rate of decrease in downwelling irradiance with increasing
aquatic depth, z, i.e.,

$$dE_d(z,\lambda)/dz = -K_d(z,\lambda)E_d(z,\lambda). \tag{3.3}$$

$K_d(z,\lambda)$ is an important parameter in light penetration models that compute
primary production as a function of available photosynthetic radiation (e.g.,
Fee, 1990). It also is used as an index of water quality in ocean color models
and is a variable that is estimable from remote measurements of ocean color.

A similar equation defines the diffuse attenuation coefficient for
upwelling irradiance, $K_u(z,\lambda)$, as the rate of decrease in upwelling irradiance
with increasing depth. Kirk (1989) has introduced a variation of this diffuse
upwelling attenuation coefficient that takes into account the depth at which
the upwelling flux was first generated. Units of radiance, L, are watts per
square meter per steradian ($Wm^{-2}sr^{-1}$). Units of irradiance, E, are watts per
square meter (Wm^{-2}). Units of attenuation coefficients are inverse meters
(m^{-1}).

Remote sensing, with its data gathering sensors, variety of viewing
directions and geometries, and vagaries of intervening atmosphere, is forced
to record the apparent properties of the water body it is monitoring.
Remotely sensed aquatic color is therefore an apparent optical property, due
to its relationship to the subsurface volume reflectance. However, the intrinsic colors of CPAs are not apparent optical properties and therefore do not
depend upon the angular and spatial distribution of the impinging radiation.
Thus, in order to utilize water color as an indicator of water quality (in terms
of the coexisting concentrations of chlorophyll, suspended minerals, and
dissolved organic carbon as discussed in Chapter One), inherent optical
properties (IOPs) of the water body are required. IOPs are those properties
of the water column independent of the spatial distribution of the impinging
radiation and the manner in which the in-water radiation field is measured.
They depend solely upon the medium being monitored. Such IOPs include:

- The absorption coefficient, $a(\lambda)$, which is defined as the fraction of
  radiant energy absorbed from an incident light beam as it traverses
  an infinitesimal distance dr divided by dr

- The scattering coefficient $b(\lambda)$, which is defined as the fraction of radiant energy scattered from an incident light beam as it traverses an infinitesimal distance dr divided by dr
- The total (or beam) attenuation coefficient, $c(\lambda)$, which is defined as the fraction of radiant energy removed from an incident light beam, due to the combined processes of absorption and scattering, as it traverses an infinitesimal distance dr divided by dr

By definition, the total light beam attenuation within a natural water body is the sum of the attenuations resulting from absorption and from scattering. Thus,

$$c(\lambda) \equiv a(\lambda) + b(\lambda). \tag{3.4}$$

Because $a(\lambda)$, $b(\lambda)$, and $c(\lambda)$ refer to the totality of attenuation of incident radiant energy due to molecular water plus all resident interstitial organic and inorganic absorption and scattering centers within the water column, they represent bulk inherent optical properties of the natural water body. Again, the units for all three attenuation coefficients are inverse meters (m^{-1}). Note that the *inherent* optical property, $c(\lambda)$, is distinct from the *apparent* optical properties, $K_d(\lambda)$ and $K_u(\lambda)$.

Scattering also results in changes in direction of motion of the scattered photons, so the spatial distribution of the scattered photons as a function of scattering angle is important. The inherent optical property volume scattering function, $\beta(\theta,\Phi)$, is defined as the scattered radiant intensity dI in a direction (θ,Φ) per unit scattering volume dV normalized to the value of the incident irradiance, E_{inc}, and provides this information. It is routinely observed that the volume scattering function possesses symmetry about the incident direction; therefore, for a constant value of scattering angle, $\beta(\theta,\Phi)$ is invariant with respect to Φ. Thus, the volume scattering function may be reduced (in units of m^{-1}sr^{-1}) to

$$\beta(\theta) = dI(\theta)/E_{inc}dV. \tag{3.5}$$

Despite the angular symmetry, however, direct determinations of $\beta(\theta)$ are not simple to perform, and only a limited number of directly determined $\beta(\theta)$ spectra are available. For many years the excellent San Diego Harbor work of Petzold (1972) has provided this valuable inherent optical property for use in selected natural waters as well as in the development of water quality models. However, a dearth of $\beta(\theta)$ spectra for natural waters is an issue that needs to be addressed.

The integral of the volume scattering function, $\beta(\theta)$, over the hemisphere trailing the incident flux, which is defined by the angular ranges ($\pi/2 \leq \theta \leq \pi$) and ($0 \leq \Phi \leq 2\pi$), yields the *backscattering coefficient*, $b_B(\lambda)$. This in turn is the product of the scattering coefficient, $b(\lambda)$, and the *backscattering probability*, $B(\lambda)$, which is defined as the ratio of the scattering into the hemisphere trailing the flux to the total scattering into all directions. Thus, $B(\lambda) = b_B(\lambda)/b(\lambda)$. For completeness, the integral of the volume scattering function $\beta(\theta)$ over the hemisphere preceding the incident flux, defined by the angular ranges ($0 \leq \theta \leq \pi/2$) and ($0 \leq \Phi \leq 2\pi$), yields the *forwardscattering coefficient*, $b_F(\lambda)$. This in turn is the product of the scattering coefficient $b(\lambda)$ and the *forwardscattering probability*, $F(\lambda)$, which is defined as the ratio of the scattering into the hemisphere preceding the incident flux to the total scattering into all directions. Thus, $F(\lambda) = b_F(\lambda)/b(\lambda)$. Clearly, the sum of $b_F(\lambda)$ and $b_B(\lambda)$ is $b(\lambda)$ and the sum of $F(\lambda)$ and $B(\lambda)$ is unity. For most scattering particles, the scattering is peaked in the forward direction (i.e., in the direction of the incident flux).

Many authors have considered the mathematical relationships linking the AOP attenuation coefficients (K_d and K_u) with the IOP attenuation coefficients (c, b_B, and a). Scattering phase function shape factors (Stavn and Weidemann, 1989; Aas, 1987, among others) are generally used to convert the backscattering coefficient, b_B, into the upward and downward irradiance attenuation coefficients, K_d and K_u.

The fundamental principles of all optical processes are incorporated within the theory of radiative transfer of energy. Because explicit mathematical formulism of the radiative transfer equation is intimidating and elusive, semiempirical relationships linking the properties of the radiation field to the inherent optical properties of the attenuating medium have been sought. The most widely used approach has been Monte Carlo simulations of photon propagation through such attenuating media.

Application of Monte Carlo analyses to ocean–atmosphere interactions was cleverly pioneered by Plass and Kattawar (1972). Since then, a number of workers — including Gordon and Brown (1973); Gordon et al. (1975); Kirk (1981a, b, 1984); Jerome et al. (1988); Stavn and Weidemann (1989); Bannister (1990); Morel and Gentili (1991, 1993); and Sathyendranath and Platt (1997, 1998) — have used Monte Carlo analyses to relate the inherent optical properties of natural waters to the apparent optical properties. Not unexpectedly, these simulations have led to quite similar, but not identical, expressions for volume reflectance as a function of the bulk inherent optical properties of the water body. Invariably, these relationships involve the bulk backscattering coefficient — the product of the backscattering probability, B, and the scattering coefficient, b, expressed in the scientific literature by $b_B(\lambda)$ or $(Bb)(\lambda)$ — and the bulk absorption coefficient $a(\lambda)$. As examples of Monte Carlo outputs, Kirk (1984) obtained for the subsurface volume reflectance, $R(0^-)$, the relationship

$$R(0^-) = (0.075 - 0.620\mu_0)b_B/a \qquad (3.6)$$

where $\mu_0 = \cos(\theta_0)$, where θ_0 is the in-water refracted angle. For overcast conditions, Equation 3.6 reduced to

$$R(0^-) = 0.437b_B/a \tag{3.7}$$

Jerome et al. (1988) obtained the relationships

$$R(0^-) = (1/\mu_0)0.319b_B/a \tag{3.8}$$

for $0 \leq b_B/a \leq 0.25$, and

$$R(0^-) = (1/\mu_0)[0.013 + 0.267b_B/a \tag{3.9}$$

for $0.25 \leq b_B/a \leq 0.50$, where $\mu_0 = 0.858$ for overcast conditions.

Equation 3.6 to Equation 3.9 hold throughout the visible wavelength spectrum. They do not, however, consider trans-spectral (i.e., inelastic) optical processes (see Section 3.5).

3.2   Case 1 and case 2 waters

The degree of optical complexity of a natural water body, in general, rapidly escalates as its proximity to a land mass increases. Thus, the relative optical simplicity of mid-ocean waters is usually (although not always) denied the waters of lakes, rivers, and coastal oceans. This optical complexity, initially based upon aquatic transmittance (defined as the ratio of the radiant flux transmitted by an attenuating medium to the incident radiant flux impinging upon it) was first proposed by Jerlov (1951, 1953). Other optical classification schemes have been proposed by Pelevin and Rutkovskaya (1977); Smith and Baker (1978); Kirk (1980); Prieur and Sathyendranath (1981); and others.

The current bipartite (case 1 and case 2) classification scheme was introduced by Morel and Prieur (1977) and refined by Gordon and Morel (1983) and others. By definition, case 1 waters are those whose optical properties are determined by molecular water along with phytoplankton and their co-varying detrital matter (i.e., chlorophyll and molecular water are principal CPAs of case 1 waters). Case 2 waters are those whose optical properties are determined by non-co-varying inorganic suspended particulates, dissolved organic matter, and possibly suspended organic matter of terrestrial origin, in addition to molecular water, phytoplankton, and detritus. That is, molecular water, chlorophyll, suspended particulates, and dissolved organic matter are principal CPAs of case 2 waters.

From these definitions, it can be seen that although most inland and coastal waters would subscribe to the case 2 criteria, some could subscribe to the case 1 criteria. Mid-ocean waters, however, would be archetypically case 1. The fact that the term "case 1" can apply to inland, coastal, and mid-ocean waters and that the term "case 2" can apply to inland, coastal, and the somewhat nebulous transition areas between coastal and mid-ocean waters is indicative of inherent weakness in such a bipartite aquatic classification. With the increasing focus on inland and coastal waters and the vast variety of CPA combinations that convolve into aquatic color, focus on the evolution of more distinctive criteria for classifying water bodies is also increased. Such activity is currently in progress. However, to be consistent with recent optical modeling, remote sensing research, and scientific jargon we will continue discussions using the case 1 and case 2 bipartite terminology (with a certain trepidation).

Within the preceding caveats, therefore, remote sensing of case 1 waters can be satisfactorily treated as a single variable problem, remote sensing of case 2 waters must be treated as a nonlinear, multivariate problem. Optical models must therefore be developed and modified accordingly.

3.3   Inherent optical properties (optical cross section spectra)

In order to extract water quality information from satellite data collected over oceans (optically deep case 1 waters), it was quickly realized that ocean color must be used to infer concentrations of the principal ocean color-producing agent: chlorophyll *a*, the generally adopted surrogate for phytoplanktonic biota. It was also quickly realized that this meant that AOP volume reflectance had to be converted into the IOPs of chlorophyll: the "specific" (as opposed to "bulk") chlorophyll absorption and backscattering coefficients pertinent to the indigenous oceanic biota. There was indeed an end-user interest in ocean water quality and, consequently, a marketplace for these products as well as funding to enable their production. The relative optical simplicity resulting from the fact that case 1 waters could be considered optically binary (in terms of principal CPAs — namely, molecular water and chlorophyll, the detritus co-varied with the chlorophyll) was an added bonus.

In an epic work, Gordon et al. (1975) used volume scattering functions from Kullenberg (1968) and curve-fitting to Monte Carlo simulations to relate the apparent optical properties $K(\lambda)$ and $R(\lambda)$ to the inherent optical properties $c(\lambda)$, $F(\lambda)$, $B(\lambda)$, and $\omega_0(\lambda)$ (the *scattering albedo* defined as the number of scattering interactions within a fixed volume of an attenuating medium expressed as a fraction of the total number of scattering and absorption interactions that occur within that fixed volume). Because $a(\lambda) + b(\lambda) = c(\lambda)$, $\omega_0(\lambda) = b(\lambda)/[a(\lambda) + b(\lambda)]$. This work led to the proliferation of Monte Carlo simulations of photon propagation resulting in bio-optical and water quality models.

In addition to IOPs being independent of the configuration of the radiative field and the manner in which that field is measured, the use of these intrinsic water properties offers a major advantage. Specific IOPs of a water column are the properties that may be attributable to a unit concentration of each individual absorption and scattering center comprising the water column. In a multicomponent aquatic medium, unlike the situation for the apparent optical properties, the *bulk inherent* optical properties may be expressed as simple summations of the individual *specific inherent* optical properties of the components of that aquatic medium. That is,

$$a(\lambda) = \sum_{1}^{n} a_i(\lambda)x_i$$

$$b(\lambda) = \sum_{1}^{n} b_i(\lambda)x_i \tag{3.10}$$

$$b_B(\lambda) = \sum_{1}^{n} (b_B)_i(\lambda)x_i$$

where

$a(\lambda)$, $b(\lambda)$, and $b_B(\lambda)$ are, respectively, the spectral bulk attenuation coefficient, spectral bulk scattering coefficient, and spectral bulk backscattering coefficient of the water column

$a_i(\lambda)$, $b_i(\lambda)$, and $(b_B)_i(\lambda)$ are, respectively, the spectral inherent attenuation coefficient, spectral inherent scattering coefficient, and spectral inherent backscattering coefficient of the ith component of the water column

The concentration of the ith component is x_i.

In limnological literature, the specific inherent optical properties are commonly referred to as *optical cross section spectra* (first used by Bukata et al., 1981a). This term is adopted from atomic and nuclear collision theory and rightfully implies that a specific aquatic component (water, chlorophyll, suspended mineral, or dissolved organic carbon) will act as an effective target for a photonic interaction (i.e., bombardment resulting in absorption or scattering of the impinging photon). Consistent with this terminology, the units of optical cross sections (as can be seen from Equation Set 3.10) are area per unit mass of the aquatic component (e.g., square meters per milligram, m^2mg^{-1}). Thus, the oceanographic spectral IOP terms *specific absorption*, *specific scattering*, and *specific backscattering coefficients* are interchangeable with the limnological terms *absorption cross section spectra*, *scattering cross section spectra*, and *backscattering cross section spectra*, respectively.

As seen from Equation Set 3.10, the optical cross section spectra (specific inherent optical properties) provide the linkages between the bulk inherent optical properties of a water body and the concentrations of its CPAs. As seen from Equation 3.6 through Equation 3.9, the bulk inherent optical properties of a water body determine its subsurface volume reflectance. The volume reflectance, in turn, is related to the water-leaving radiance and, therefore, remotely sensed radiometric water color. Thus, the optical cross section spectra (spectral IOPs) provide the linkages between water color and water quality.

3.4   Forward and inverse optical modeling

From Equation Set 3.10, it is seen that knowledge of the coextant concentrations of aquatic CPAs and their optical cross section spectra (IOPs) makes it possible to construct the bulk inherent optical properties of the water body being studied. Through relationships such as those of Equation 3.6 through Equation 3.9, these bulk optical properties may be used to model the aquatic color resulting from a variety of possible combinations of CPAs. In the context of remote sensing, this generation of volume reflectance, $R(0^-,\lambda)$, or water-leaving radiance, $L_u(0^+,\lambda)$ (Equation 3.1 and Equation 3.2) from knowledge of the optical cross section spectra of the indigenous aquatic components is referred to as *forward optical modeling*.

In the context of remote sensing, *inverse optical modeling* deconvolves water color into the concentrations of organic and inorganic colorants that have collectively generated that water color. Equation Set 3.10 could arguably represent either forward or inverse modeling. Knowledge of CPA concentrations (from *in situ* water samples laboratory analyzed for composition) and bulk optical properties (from *in situ* optical measurements in tandem with sample collection) could, through multivariate optimization modeling techniques, yield the optical cross section spectra. Similarly, knowledge of the optical cross section spectra and the bulk optical properties could — again through multivariate optimization modeling techniques — yield the coexisting CPA concentrations.

In the partitioning of the bulk inherent properties $a(\lambda)$, $b(\lambda)$, and $b_B(\lambda)$ defined in Equation Set 3.10, constant values of $a_w(\lambda)$, $b_w(\lambda)$, and $(b_B)_w(\lambda)$ are considered the optical cross section values for molecular water. Thus, because case 1 waters may be considered as an admixture of water and chlorophyll and its mortality-related detritus, the n in the summation equations (Equation 3.10) is 1. This fortuitous mid-ocean condition has enabled chlorophyll concentrations to be remotely inferred from optical data using a variety of empirical algorithms (e.g., Gordon and Clark, 1980; Gordon et al., 1983; Morel, 1980; Smith and Wilson, 1981) based upon ratios of mathematically manipulated upwelling spectral radiance values, $L(\lambda)$, recorded at satellite or aircraft altitudes. Such algorithms assume the form

$$C_{chl}=N\left[\frac{L(\lambda_1)}{L(\lambda_2)}\right]\exp(M) \tag{3.11}$$

where C_{chl} is the chlorophyll concentration; $L(\lambda_1)$ and $L(\lambda_2)$ are the upwelling radiances at two different wavelengths λ_1 and λ_2; and N and M are empirically determined constants. Detailed discussions of these empirical algorithms may be found in Sathyendranath and Morel (1983).

These case 1 chlorophyll retrieval algorithms, however, are derived solely from regression techniques and thus ignore the specific IOPs of the remotely sensed water body. For case 1 waters, the principal IOPs (apart from those of molecular water) would be the absorption and scattering cross section spectra of the indigenous chlorophyll bearing biota. Interestingly, empirical relationships have been developed and applied (e.g., Lee et al., 1996a, b) to determine ocean IOPs from ratios of spectral remote sensing reflectance.

For case 2 waters, the increased number of color-producing agents and the greater ranges in their concentrations would prohibit the use of such empirical oceanic chlorophyll retrieval algorithms. Consequently, in their study of Lake Ontario, Bukata et al. (1981a, b) devised a four-component model for inland waters. This model was based upon the absorption cross section and the backscattering cross section spectra of molecular water [a_w, $(b_B)_w$]; chlorophyll a [a_{chl}, $(b_B)_{chl}$]; suspended mineral [a_{sm}, $(b_B)_{sm}$]; and dissolved organic carbon [a_{DOC}, $(b_B)_{DOC}$]. Thus, for inland and coastal waters, the relationships between bulk and specific inherent optical properties become

$$a(\lambda) = a_w(\lambda) + a_{chl}(\lambda)x_{chl} + a_{sm}(\lambda)x_{sm} + a_{DOC}(\lambda)x_{DOC} \qquad (3.12a)$$

and

$$(b_B)(\lambda) = (b_B)_w(\lambda) + (b_B)_{chl}(\lambda)x_{chl} + (b_B)_{sm}(\lambda)x_{sm} + (b_B)_{DOC}(\lambda)x_{DOC} \qquad (3.12b)$$

where x_{chl}, x_{sm}, and x_{DOC} are the coextant chlorophyll, suspended mineral, and dissolved organic carbon concentrations, respectively. A similar equation may be written for the bulk inherent scattering cross section, $b(\lambda)$. Because scattering from dissolved organic matter is generally assumed negligible, the backscattering term, [$(b_B)_{DOC}(\lambda)x_{DOC}$], is invariably set to zero.

Again, it is evident from Equation Set 3.12a, b that, if the bulk inherent optical properties of the water body, [$a(\lambda)$ and $(b_B)(\lambda)$], are known in concert with the CPA concentrations, [x_{chl}, x_{sm}, and x_{DOC}], then multivariate optimization analyses can yield the spectral values of optical cross sections, [$a_{chl}(\lambda)$, $a_{sm}(\lambda)$, $a_{DOC}(\lambda)$, $(b_B)_{chl}(\lambda)$, $(b_B)_{sm}(\lambda)$], pertinent to the natural water body. Similarly, if the bulk inherent optical properties of the water body are known in concert with the pertinent optical cross section spectra, the coexisting concentrations of the indigenous CPAs may be determined. A popular multivariate analyses technique is attributed to Levenberg (1944) and Marquardt (1963); this has been shown to provide very good agreement between directly measured and remotely inferred values of chlorophyll a, suspended minerals, and dissolved organic carbon (Bukata et al., 1985; Pozdnyakov et al., 1999) for the optically complex waters of the Laurentian Great Lakes.

A variety of inverse modeling techniques and algorithms have been and are being developed to retrieve chlorophyll concentrations from remote measurements of water color. The simple determination of ratios (or other mathematical manipulation) of satellite radiances (Equation 3.11) is one such family of algorithms that works surprisingly well for most case 1 waters. That such simple manipulations do not hold for case 2 waters is a consequence of the terrestrial-driven organic and inorganic matter that optically compete with in-water chlorophyll as dispersion centers for the downwelling global radiation.

With a subsequent focus shift from mid-ocean to coastal ocean remote sensing, understandable, although misdirected, attempts were made to "tweak" algorithms that were ably serving mid-ocean requirements. The hope was that such tweaking would enable these algorithms to serve coastal water requirements ably. Scientists dealing with the complexities of limno-logical optics, however, had quickly realized that such tweaking would be inappropriate and that, conversely, algorithms developed for case 2 waters could, in their simple limits, be applicable to case 1 waters.

Further complicating the extraction of water quality information from the color of inland and coastal waters were the facts that the *same* aquatic color could result from *different* combinations of *chl*, *sm*, and *DOC*, and that *different* aquatic colors could result from the *same* combinations of *chl*, *sm*, and *DOC*. Consequently, it became apparent (Bukata et al., 1981b) that, in order to extract the concentration of a desired CPA (e.g., *chl*), the coextant concentrations of all other CPAs (e.g., *sm*, *DOC*) would have to be extracted simultaneously. This requires knowledge of the optical cross section spectra of the site-specific and time-dependent indigenous organic and inorganic CPAs.

Thus, multivariate analyses such as Levenberg–Marquardt optimization or simplex algorithm (Nelder and Mead, 1965) form major components of inverse modeling of remotely sensed data acquired over inland and coastal waters. Other multivariate inverse retrieval approaches include

- Principal component inversion (Mueller, 1976; Fischer, 1985; Sathy-endranath et al., 1994)
- Neural networks in which piecewise-linear mapping is achieved be-tween a set of spectral radiances impinging upon the top of the atmosphere and a selected set of water quality variables (Keiner and Yan, 1998; Doerffer and Schiller, 1998, 1999)
- Quantum adiabatic analyses in which a succession of quantum me-chanical operations is utilized (Steffen et al., 2003; Lee et al., 1996a, b)

In addition to top-of-the-atmosphere radiances, these approaches require input values of the optical properties of the atmospheric and aquatic con-stituents, the state and properties of the air–water interface, and the viewing conditions and orientation of the satellite or airborne sensor. The complex-ities, applications, and validations of these multivariate approaches will not be pursued here. Suffice to say that algorithms directed towards the extrac-tion of case 2 water quality variables from remote measurements of aquatic color incorporate such multivariate analyses. The reader is urged to become familiar with these and other multivariate approaches to analyzing remotely acquired spectro-optical data over aquatic environmental targets, as well as over terrestrial (including inland and coastal wetlands) environmental targets.

Throughout these discussions, the specific spectral IOPs of aquatic con-stituents have been emphasized. The optical cross section spectra of molec-ular water are available in the scientific literature. The spectra determined by Smith and Baker (1981) that have served the optical community as a standard for two decades have been updated by Pope and Fry (1997). The sharp focus on remotely monitoring mid-ocean waters has resulted in an abundance of *in situ* and laboratory determinations of the absorption cross section spectra of a variety of oceanic chlorophyll-bearing biota (e.g., species of red algae, green algae, blue-green algae). Considerably less emphasis has been placed on the scattering cross section spectra of these phytoplankton, perhaps because determinations that have been performed (e.g., Bricaud et al., 1983) indicate very low values for backscattering coefficients of phy-toplankton. There has also been an historical focus on the optical properties of dissolved organic matter, the so-called *yellow substances*. These are defined as water-soluble polymers or humus that imparts a yellow hue to natural waters. Yellow substances are also referred to as *gelbstoff, aquatic humic matter, yellow organic acids, gilvin,* and *humolimnic acid.*

There is excellent agreement in the spectral form of absorption cross sections of dissolved organic matter as reported in the scientific literature (e.g., Unoki et al., 1978; Bricaud et al., 1981; Bukata et al., 1981b; Roesler et al., 1989; Gallegos et al., 1990): namely, a distinctive exponential decrease with increasing wavelength throughout the visible spectrum. Variability in the slope of this exponential decrease is explained (e.g., Carder et al., 1989; Kopelevich and Ershova, 1997; Højerslev, 1998) by variations in the compo-sition of the yellow substances. In this book, we consider dissolved organic carbon (DOC) to be a reasonable surrogate for dissolved organic matter (yellow substances), a practice consistent with that of a large segment of the aquatic optics community.

The greatest discrepancies in optical cross section spectra of case 2 waters are in those pertinent to suspended inorganic particulate matter of terrestrial origin. These discrepancies are attributable to geologic diversities. The sus-pended inorganic particulate concentration of local origin has become regarded as merely an irritating backscatter that could be "corrected" out of the saga of case 2 water science, perhaps due to (1) a long-standing preoc-cupation with ocean optics; (2) a concurrent and enigmatic long-standing lack of recognition of limnological optics; (3) an all too prevalent philosophy that a simple tweak could render case 1 water quality algorithms apropos to case 2 water quality determinations; or (4) all of these reasons. As seen from Equation Set 3.12a, b, suspended inorganic particulates (or, as consid-ered here, the quantifiable surrogate, suspended minerals) are independent color-producing agents (CPAs) of the water column. Some inverse models employed to infer estimates of case 2 water chlorophyll consider the non-chlorophyllous pigment CPA as additive to organic detrital matter and fur-ther consider a composite backscattering term to account for turbidity.

Such considerations fail to respect the independence of terrigenous suspended inorganic particulates as CPAs that are often the principal determinant of local aquatic color. To date, the number of directly measured optical cross section spectra for inland and coastal waters has been extremely limited; those that exist are far from identical in spectral shape and spectral intensity (e.g., Bukata et al., 1985, 1991a, b; Gallie and Murtha, 1992; Morel and Prieur, 1977; Prieur and Sathyendranath, 1981; Whitlock et al., 1981). Happily, however, it appears that this issue is being addressed. Perhaps suspended inorganic matter will be seriously considered, at last, as a dominant CPA of inland waters (see Section 6.6).

Our knowledge of the optical cross section spectra of suspended inorganic particulates of terrestrial origin remains meager, and this is a matter that must be addressed. The geologic composition of a basin plays a major role in the color of its adjacent or encapsulated waters, in terms of indigenous suspended and dissolved inorganic and organic CPAs. Because the greatest discrepancies in the optical cross section spectra pertinent to indigenous inland water CPAs are consequences of global geologic diversities, it would be ideal to determine and catalogue these optical properties directly for every case 2 national and international water body. Indeed such a catalogue has been advocated for some time, although a relatively sparse number of entries comprise this catalogue to date.

Direct determinations of such specific inherent optical properties (as seen from Equation Set 3.12a, b) are time-, labor-, and cost-intensive, requiring, as they do, a combination of *in situ* optical measurements in concert with water sampling followed by laboratory analyses and multivariate inverse modeling. The use of surrogate optical cross section spectra, if possible, would be of great benefit to forward and inverse modeling. Despite the limited number of entries in the global IOP catalogue, certain similarities among these properties have been observed, particularly for the optical cross section spectra of freshwater biota (even though dissimilar from those of saltwater biota) and for freshwater dissolved organic matter.

Suspended particulates remain a problem. Recently, Bukata et al. (2001a) utilized optical measurements and water quality data from inland waters in Canada, Russia, the U.S., and Switzerland and indicated that compensation for geologic diversities *may* be possible by judicious selection of watersheds displaying comparable geologic origins and/or histories. An ability of waters of geologic compatibility to use common (or at least similar) optical properties could alleviate the time-, labor-, and cost-intensive determinations of these properties for many inland water bodies. Such a possibility is still far from becoming a global reality; nevertheless, it is an important possibility to consider when arriving at a site-dependent remote sensing compromise between defendable science and defendable usage of that science (see Section 1.8).

3.5 Trans-spectral processes

Until this point in our discussions, we have been considering spectral processes in which scattered photons, while undergoing attenuation of energy and changes in direction of propagation, do not undergo changes in wavelength or polarization. Wavelength-invariant scattering processes define *elastic scattering*. However, *inelastic scattering* is defined as scattering in which scattered photons undergo changes in wavelength and polarization. Raman scattering within natural waters is an inelastic scattering process. Stimulated *fluorescence* (defined as the release of energy at one wavelength by a substance subsequent to its absorption of energy at some other wavelength) is another inelastic process. Fluorescence may result from stimulation of phytoplanktonic pigments or dissolved organic matter. Raman scattering and fluorescence are termed *trans-spectral processes* because stimulation at wavelength λ_1 leads to emission at wavelength λ_2 where, in general, $\lambda_2 > \lambda_1$.

Perhaps due to the optical complexity of case 2 waters, optical and bio–geo-optical models developed for the extraction of water quality variables from the color of inland and coastal waters have usually considered only the effects of absorption and elastic scattering *per se* on the subsurface volume reflectance or water-leaving radiance spectrum (or any other indicator of water color). However, in addition to absorption and scattering, Raman scattering and fluorescence are known to have an impact on the spectral distribution of upwelling radiance in case 1 waters (Sugihara et al., 1984), thereby modifying sea-water color and affecting bio-optical models (Peacock et al., 1990).

Therefore, the subsurface volume reflectance, $R(0^-,\lambda)$, of case 1 and case 2 waters may be somewhat simplistically considered as the additive consequences of the spectral and trans-spectral energy transfer processes occurring within the water column. At least for single scattering, $R(0^-,\lambda)$ may be compartmentalized as:

$$R(0^-,\lambda) = R_{es}(0^-,\lambda) + R_r(0^-,\lambda) + R^f_{chl}(0^-,\lambda) + R^f_{DOC}(0^-,\lambda) \qquad (3.13)$$

where

$R_{es}(0^-,\lambda)$ is the contribution of absorption and elastic scattering

$R_r(0^-,\lambda)$ is the contribution of inelastic Raman scattering

$R^f_{chl}(0^-,\lambda)$ is the contribution of trans-spectral fluorescence from chlorophyll

$R^f_{DOC}(0^-,\lambda)$ is the contribution of trans-spectral fluorescence from dissolved organic carbon

In Equation 3.13, $R_{es}(0^-,\lambda)$ values are usually (but as we shall discuss shortly, perhaps inappropriately) considered as those obtained from the $R(0^-)$ equations (Equation 3.6 to Equation 3.9).

Trans-spectral processes are characterized by nearly isotropic phase functions, $P(\theta)$, which are defined as the volume scattering function, $\beta(\theta)$, normalized to the scattering coefficient, $b(\lambda)$, and calculated as $4\pi\beta(\theta)/b(\lambda)$. Thus, Raman scattering and stimulated fluorescence result in multiple downward scattering events that may discredit the single-scattering $R(0^-,\lambda)$ compartmentalization of Equation 3.13. Sathyendranath and Platt (1998) separate first-order Raman scattering events from Raman–elastic and Raman–Raman events and show that the contribution to Raman scattering by photons undergoing second-order Raman scattering would only be about 10% that of first-order Raman scattering. Each successive order would add a further 10% of the prior order. It would not be unreasonable to reckon that similar multiple scattering relationships would hold for fluorescence from chlorophyll and dissolved organic carbon.

We will not pursue an in-depth treatise on the mathematical formulism of the radiative processes governing trans-spectral processes occurring in natural waters here. Rather, the reader is referred to the excellent work in the literature, a woefully incomplete list of which is cited within this section.

Sathyendranath and Platt (1998) mathematically express $R_r(0^-,\lambda)$ as a function of the Raman backscattering coefficient, $(b_B)_r$; the attenuation coefficients for downwelling and upwelling irradiances, $K_d(\lambda)$ and $K_u(\lambda)$; the downwelling irradiance just beneath the air–water interface, $E_d(0^-,\lambda)$; the average cosines for downwelling and upwelling irradiances, μ_d and μ_u; and the Raman scattering excitation and emission wavelengths, λ_{ex} and λ_{em}.

Incorporating fluorescence into radiative transfer calculations requires knowledge of the fluorescence quantum yield (a.k.a. fluorescence quantum efficiency), $\eta(\lambda_{ex},\lambda_{em})$, of the fluorescing matter (i.e., the ratio of the number of photons of emission wavelength λ_{em} to the number of absorbing photons of wavelength λ_{ex}). Although precise $\eta(\lambda_{ex},\lambda_{em})$ values are unavailable for many specific species of fluorescing matter, typical values of η for chlorophyll, η_{chl}, generally range from 0.005 to 0.05; typical values of η for dissolved organic matter (e.g., carbon), η_{DOC}, generally range from 0.005 to 0.03 (IOCCG, 2000). Shape factors of η_{chl} and η_{DOC} are discussed by Gordon (1979), Hawes (1992), Culver and Perry (1997), and others. The chlorophyll fluorescence emission peak occurs at 685 nm and the DOC fluorescence emission peak is in the range 490 to 530 nm. Incorporating fluorescence into forward modeling of volume reflectance is further complicated by the fact that the same compounds that produce the fluorescence also act as quenching agents of that fluorescence.

Gordon (1979) mathematically expresses $R^f_{chl}(0^-,\lambda)$ as a function of the chlorophyll absorption coefficient, $a_{chl}(\lambda_{ex})$, at the excitation wavelength, λ_{ex}; the chlorophyll fluorescence quantum yield, $\eta_{chl}(\lambda_{ex},\lambda_{em})$; the dispersal of the chlorophyll emission band about its peak emission wavelength; and the downwelling irradiance attenuation coefficients, $K_d(\lambda_{ex})$ and $K_d(\lambda_{em})$, at the excitation and the emission wavelengths.

Culver and Perry (1997) mathematically express $R^f_{DOC}(0^-,\lambda)$ as a function of the DOC absorption coefficient, $a_{DOC}(\lambda_{ex})$, at the excitation wavelength, λ_{ex}; the DOC fluorescence quantum yield, $\eta_{DOC}(\lambda_{ex},\lambda_{em})$; the dispersal of the DOC emission band about its peak emission wavelength; and the downwelling irradiance attenuation coefficients, $K_d(\lambda_{ex})$ and $K_d(\lambda_{em})$, at the excitation and the emission wavelengths.

Additional information on trans-spectral processes may be found in Kishino et al. (1986); Marshall and Smith (1990); Fischer and Kronfield (1990), Haltrin and Kattawar (1993), Bartlett et al. (1998); Green and Blough (1994); Coble and Brophy (1996); Vodacek et al. (1994); Babin et al. (1996); and Loisel and Stramski (2000), among many others.

Some quick and pertinent results emerging from the works cited in this section would include:

- Maximum impacts of in-water Raman scattering would logically occur in water devoid of chlorophyll, suspended minerals, and dissolved organic matter. Because suspended organic and inorganic matter are the principal participants in absorption and elastic scattering processes in inland and coastal waters, the inelastic Raman scattering is most prominent in oligotrophic (mid-ocean and some inland) waters. Raman scattering would be of little or no consequence to the water-leaving radiance of case 2 waters containing even relatively small concentrations of elastic scatterers.

- Although Raman scattering decreases with increasing wavelength, its *percentage* impact on pure molecular water increases with increasing wavelength throughout the visible spectrum. Maximum impacts occur at $\lambda > $ ~635 nm, where Raman scattering accounts for nearly 35% of the volume reflectance at $\lambda \sim 685$ nm. (However, this value of $R(0^-,685)$ is $< 2 \times 10^{-4}$. By contrast, for water containing no suspended inorganic or dissolved organic matter, but a chlorophyll concentration of 1.0 μg/L, the value of $R(0^-,685)$ is $\sim 6 \times 10^{-2}$ with fluorescence accounting for over 99% of this reflectance.)

- Fluorescence is most prominent in optically binary (i.e., molecular water plus one of chlorophyll or of DOC) mesotrophic waters. This prominence will increase as the concentration of the fluorescing matter increases to some critical concentration. Again, fluorescence would be of consequence to the water-leaving radiance of mid-ocean and some inland waters.

- The impact of chlorophyll fluorescence on volume reflectance (and water-leaving radiance) manifests as a well-defined Gaussian distribution around λ_{em} (685 nm). The impact of DOC fluorescence on volume reflectance (and water-leaving radiance) manifests as an ill-defined broadband enhancement over an extended spectral range from ~430 to ~660 nm.

- Due to the fluorescence-quenching aspects of chlorophyll and DOC in addition to the elastic scattering by suspended particulates, prominence of chlorophyll-induced and/or DOC-induced fluorescence will dramatically decrease as case 2 waters become more turbid.

Thus, impacts of Raman scattering on the spectral distribution of water-leaving radiance would be readily detectable in waters devoid of elastic scatterers. Impacts of fluorescence on the spectral distribution of water-leaving radiance

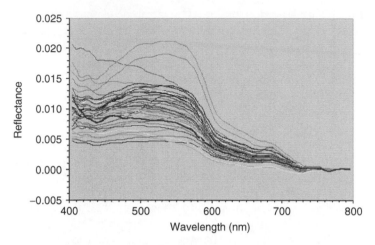

Figure 3.1 Reflectance spectra recorded by the Institute of Ocean Sciences spectrometer in and around Saanich Inlet. (From Bukata et al., *Can. J. Remote Sens.,* 30, 8–16, 2004.)

would be readily detectable in waters rich in phytoplankton or dissolved organics but containing only small to moderate amounts of suspended minerals.

Figure 3.1, taken from Figure 1 of Bukata et al. (2004), illustrates reflectance spectra recorded by the Fisheries & Ocean Canada Institute of Ocean Science spectrometer in and around Saanich Inlet, a temperate marine fjord on Vancouver Island, British Columbia. Clearly seen is the Gaussian distribution centered on the chlorophyll fluorescence emission wavelength, λ_{em0} = 685 nm.

Figure 3.2, taken from Figure 2 of Bukata et al. (2004), illustrates the remote sensing reflectance (ratio of the water-leaving spectral radiance to

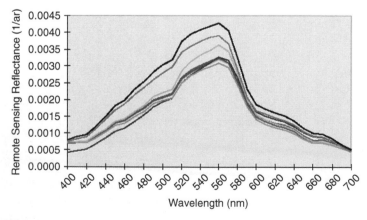

Figure 3.2 Remote sensing reflectance spectra recorded by the National Water Research Institute WATERS system of upwelling and downwelling spectrometers in Lake Waskesiu. (From Bukata et al., *Can. J. Remote Sens.,* 30, 8–16, 2004.)

the downwelling spectral irradiance; see Section 3.6) recorded by the National Water Research Institute WATERS configuration of upwelling and downwelling spectrometers in Lake Waskesiu, Saskatchewan, during the Boreal Ecosystem–Atmosphere Study (BOREAS). No obvious chlorophyll peak is discernible. As mentioned, however, discernible chlorophyll fluorescence may be observed on water-leaving radiance spectra over inland waters (e.g., some eutrophic pelagic Great Lake waters). However, in most instances, the fluorescence peak is indiscernible, minuscule, or absent.

Grassl et al. (2002) consider the impact on inland water quality retrieval that would result from the use of multivariate analyses that neglected in-water trans-spectral processes (i.e., by using equations like Equation 3.6 through Equation 3.9). Optical cross section spectra with preselected concentrations of *chl*, *sm*, and *DOC* were used to forward model $R_{es}(0^-,\lambda)$. Mathematically modeled values of $R_r(0^-,\lambda)$, $R^f_{chl}(0^-,\lambda)$, and $R^f_{DOC}(0^-,\lambda)$ were then added to obtain values of $R(0^-,\lambda)$, per Equation 3.13. The optical cross section spectra were then used to inversely model the concentrations of *chl*, *sm*, and *DOC* that would produce this value of $R(0^-,\lambda)$. These concentrations were then compared with the preselected concentrations used to forward model $R_{es}(0^-,\lambda)$. In this manner, they illustrated that, for inland and coastal waters containing admixtures of *chl*, *sm*, and *DOC*, retrieval errors for extracting concentrations *chl*, *sm*, and *DOC* from volume reflectance arising from neglect of trans-spectral processes could exceed ±300%. (The reported estimated retrieval errors were generally contained within the range ±100%.)

Such substantial uncertainties in the retrieval of organic and inorganic colorant concentrations from inland water color without due deliberation given to trans-spectral processes are startling and discouraging. Certainly, such large retrieval errors would appear to belie the quick and pertinent results emergent from the works cited in this section. If we accept that the reported uncertainties are indeed the result of trans-spectral neglect, then a focus on the reality of the modeling exercise (in particular the validity of Equation 3.13 and the numerical values of the variables that feed it) would be in order.

Forward modeling the additive elastic plus inelastic Equation 3.13 requires the implicit assumption that the addition of a CPA is not counterproductive to the impacts of inelastic contributions. The chlorophyll and dissolved organic matter trans-spectral contributions, $R^f_{chl}(0^-,\lambda)$ and $R^f_{DOC}(0^-,\lambda)$, to water color, $R(0^-,\lambda)$, should be *net* rather than *gross* values of the stimulated fluorescence. Properly taking into account the quenching and the attenuation of fluorescence within a subsurface optical field by variable admixtures of optically interactive organic and inorganic matter (although possibly a time-consuming activity of debatable accuracy) should significantly reduce the uncertainties limit.

Regarding numerical values that feed Equation 3.13 and all equations that involve inherent optical properties of natural water bodies, Bukata et al. (2004) suggest a *Gedanken* thought process. Optical cross section spectra (inherent optical properties) specific to a water body are determined and reported in two ways:

- From accurate laboratory determinations of the absorption and scattering properties of individual species aquatic matter, although rarely for suspended particulates. Here, the identity of the aquatic component, the nature of its containment medium, and the configuration of the light field are known. Determinations are conducted within precisely controlled laboratory protocols.
- From field excursions in which *in situ* optical determinations of subsurface volume reflectance, irradiance attenuation, and total attenuation coefficients are performed in tandem with the collection of water samples brought back to the laboratory for accurate determinations of *chl*, *sm*, and *DOC* concentrations.

As we have seen, different sets of indigenous inherent optical properties would produce different values of $R(0^-,\lambda)$. So too, however, might the sets of inherent optical properties emerging from the laboratory and field determinations at the same time for the same water body. Invariably, direct determinations of IOPs under controlled laboratory conditions are the results of absorption and elastic scattering. However, *in situ* determinations of subsurface volume reflectance, $R(0^-,\lambda)$, and water-leaving radiance, $L_u(0^+,\lambda)$, already include the local values of $R_r(0^-,\lambda)$, $R^f_{chl}(0^-,\lambda)$, and $R^f_{DOC}(0^-,\lambda)$.

Therefore, inverse modeling of in-water volume reflectance and known concentrations of coexisting CPAs do not yield the specific IOPs of solely absorption and inelastic scattering, but rather some hybrid of specific IOPs representing combined contributions of elastic (spectral) and inelastic (trans-spectral) processes attributable to the indigenous *chl*, *sm*, and *DOC*. (In this regard, the limnological choice of "optical cross section spectra" as an appellation for IOPs is particularly appropriate, rightfully implying that an aquatic component (water molecule, suspended particulate, dissolved or suspended organic matter) acts as a target for photonic bombardment resulting in absorption, elastic scattering, inelastic scattering, or fluorescence.)

The optical cross section spectra used in the Grassl et al. (2002) work were obtained from *in situ* expeditions of Russian lakes. In fact, the optical cross section spectra used in several North American, South American, and European lake studies are usually determined from direct observations of the targeted water columns. Thus, even if Equation 3.13 did, in effect, sum terms that were completely independent of one another (i.e., the presence of one CPA did not affect the optical behavior of a competitive CPA or its own optical behavior in any other than a mathematically precise and known manner), the impacts of trans-spectral processes become double (or higher) entrants if field determinations of inherent optical properties are used. Therefore, the ±100% is a figure that would certainly appear to be very highly inflated and one that would significantly drop with the use of field-determine optical cross section spectra.

This differential in laboratory-determined and field-determined environmental properties is, of course, not restricted to aquatic optics. Rather, it is systemic to the determination of all environmental properties resulting from ecosystems' cumulative and synergistic responses to multiple interactive stressors. However, it would appear that to minimize discrepancies between remotely inferred and directly measured water quality parameters, it is essential to utilize optical cross section spectra determined from inverse multivariate analyses of *in situ* spectro-optical data in waters of known concentrations.

We are not suggesting that bio–geo-optical models incorporating fluorescence and Raman scattering should be abandoned. Mathematical rigor is a valued commodity, and incorporation of trans-spectral processes is essential for full-color analyses of optically binary natural waters (case 1 waters that are binary in chlorophyll and case 2 waters that are binary in dissolved organic matter). The equations relating trans-spectral processes to water color are far from trivial and require precise input values, which are not always readily available, of a large number of inelastic parameters. Also, inversion of such mathematically complex equations can quickly become cumbersome and possibly insoluble. We are suggesting, however, that the determination and use of optical cross section spectra acquired *in situ* is a practical means of minimizing these difficulties for optically complex water bodies. (We are, of course, well aware that use of *in situ*-determined optical cross section spectra is not free from criticism, in addition to being labor-, time-, and cost-intensive.)

3.6 Remote sensing reflectance

In this chapter, we have emphasized the relationships that have linked the optical properties and coextant concentrations of in-water absorbing and scattering matter (CPAs) with the volume reflectance, $R(0^-,\lambda)$. Historically, $R(0^-,\lambda)$ has been the focus of bio-optical models to estimate concentrations of CPAs (Bukata et al., 1981b, 1991a; Carder et al., 1991; Jaquet et al., 1994, among others). Because satellites generally view downward at some nadir angle, θ_A, and record the upwelling radiance within an angular cone surrounding that nadir angle, complications arise when attempting to utilize a bio–geo-optical model based upon volume reflectance (ratio of upwelling to downwelling irradiances) to this upwelling radiance spectrum (not the least of which is due to the solar zenith angle dependence of the Q-factor of Equation 3.2). These complications have probably nurtured the development of empirical relationships (Equation 3.11) involving ratios of upwelling radiances at different wavelengths that worked so surprisingly well for near-surface mid-ocean chlorophyll concentration estimates from the coastal zone color scanner (CZCS) on Nimbus-7.

Some authors (e.g., Morel and Gentili, 1993; Jerome et al., 1996; Gould and Arnone, 1997) have considered remote sensing reflectance as the appropriate parameter upon which to focus in order to infer CPA concentrations from color spectra remotely measured above case 2 waters. Remote sensing reflectance, $R_{RS}(\theta_A)$, is defined (Gordon et al., 1984) as the ratio of the upwelling radiance from a given nadir direction to the downwelling irradiance; both are measured at the remote platform. A major advantage of its use in bio–geo-optical modeling is elimination of the need to estimate the troublesome Q-factor.

Neglecting the issues of sun glitter and subsurface internal reflection of in-water upwelling irradiance, the remote sensing reflectance just above the air–water interface, $R_{RS}(\theta_A)$, can be related to the remote sensing reflectance just beneath the air–water interface, $R_{RSW}(\theta_W)$, by

$$R_{RS}(\theta_A) = R_{RSW}(\theta_W)[(1 - \rho_{irr}(\theta_A)(1 - \rho(\theta_W))/n^2] \qquad (3.14)$$

where

θ_W is the in-water nadir angle

$\rho_{irr}(\theta_A)$ is the surface reflectivity for downwelling irradiance in air

$\rho(\theta_W)$ is the internal surface reflectivity for an incident angle of θ_W

n is the relative index of refraction (~1.333)

Using Monte Carlo simulations of photon propagation, Jerome et al. (1996) express the bulk inherent optical property, b_B/a, at any wavelength, as a third-order polynomial expansion of the remote sensing reflectance just beneath the air–water interface, R_{RSW} namely:

$$b_B/a = 0.0027 + 9.87R_{RSW} - 34.5\,(R_{RSW})^2 + 1534(R_{RSW})^3 \qquad (3.15)$$

The functional relationship of Equation 3.15 is essentially independent of solar zenith angle and is appropriate within the visible wavelength range for the values of $B(\lambda)$, $b(\lambda)$, and $a(\lambda)$ found in most case 2 waters (with the possible exclusion of the most highly turbid).

At any wavelength, the bulk inherent property, b_B/a, can be expanded as additive consequences of the concentrations and optical cross section spectra of the local CPAs (Equation Set 3.12a, b). Consequently, the remote sensing reflectance just above the air–water interface, $R_{RS}(\theta_A)$, serves as an effective variable to forward model optical cross section spectra and CPA concentrations into water color and to inversely model water color into the concentrations of the coexisting CPAs if the pertinent optical cross section spectra are known. Conversely, it can inversely model water color into the pertinent optical cross section spectra if the coexisting CPA concentrations are known. Reiterating from Section 3.5, in order to incorporate impacts of trans-spectral processes into the modeled optical cross section spectra, we recommend that *in situ* determinations of optical parameters be conducted in concert with collection of water samples for accurate laboratory determination of CPA concentrations.

Obviously, water quality products of great importance to environmental decision-makers and policy-makers are near-surface chlorophyll "maps," and the remote sensing scientific community should continue to strive for inclusion of such maps into the protocols of environmental monitoring networks. However, a case could be made that (at least) scientific end-users could find value in the generation of maps of aquatic optical properties (e.g., maps of the surficial values of b_B/a resulting from the remote sensing reflectance Equation 3.14 and Equation 3.15). Such maps could provide a vehicle for the much needed validation (or, at minimum, an intercomparison) of competing models, algorithms, and methodologies for the extraction of water quality information from remotely acquired data. They could also provide insight into the virtues and limitations of the varied approaches used for combating atmospheric intervention (possibly the greatest single unresolved issue plaguing virtually every Earth observation space program; see Section 3.7).

In a report resulting from the Canadian Space Agency's Earth Observation Applications Development Program (CSA's EOADP), Borstad et al. (2002a) describe a manner in which remote sensing reflectance (which is related to volume reflectance) might be used in conjunction with a regional color-based bio–geo-optical model and multivariate analyses to obtain surficial maps of aquatic water quality. Equation 3.12a, b and Equation 3.14 are combined with the Kirk (1984) and the Phillips and Kirk (1984) relationships among the irradiance attenuation coefficient, $k(\lambda)$, solar zenith angle, θ, and the bulk optical properties $a(\lambda)$, $b(\lambda)$, and $c(\lambda)$; a matrix may then be derived for each sampling station that represents a single equation with only the three CPA concentrations (chlorophyll, suspended sediments, dissolved organic matter) at that station as the unknowns. Thus, if measurements at that station are performed at N wavelengths, this equation then represents a matrix of N equations with the three CPA concentrations at that station as the unknowns. Solving these N equations through multivariate optimization yields the coextant CPA concentrations.

3.7 The status of remote sensing science

This chapter has presented a very brief run-through of some of the pertinent scientific background for remotely monitoring the quality of natural waters rendered optically complex by their proximity to or encapsulation by land masses that dominate their optical properties and, thereby, their color (the so-called case 2 waters). Because of its brevity, an enormous amount of impressive and dedicated work on a variety of inter-related topics in aquatic optics and bio-optical modeling has been omitted. We apologize for the omission and refer the reader to the excellent descriptive and instructional material available in the historical books and treatises on ocean optics referenced earlier (e.g., Preisendorfer, 1976; Jerlov, 1976; Kirk, 1983; 1994; and

Maul, 1985). Books, treatises, and reviews of the physics, biophysics, and spectro-optical remote sensing of case 2 waters are also available (including, but not restricted to, Dekker, 1993; Mobley, 1994; Bukata et al., 1995; Lindell et al., 1999; Lillesand and Kiefer, 2000; IOCCG, 2000; Dekker et al., 2001c, 2002a; Dekker and Bukata, 2002; Richardson et al., 2002).

Belying the brevity of this précis on remote sensing science, the remote sensing science is impressive, robust, and applicable to the generation of water quality products at all four levels (see Section 1.6). The interaction of photons with the air–water interface, the in-water aquatic matter, and the ensuing propagation of attenuated and/or stimulated photons through the interface and the water column is well understood and the mathematical formulism and simulation of photon behavior is readily defendable. As discussed in Chapter One, however, perverse behavior of the environment resulting from the interdependencies of its forcing functions, biogeochemical vagaries, and diversity of ecosystem membership invariably prohibits a universal application of remote sensing algorithms and methodologies that this science has produced.

Remote monitoring of ecosystem change in general and case 2 water quality in particular requires local algorithms. Such algorithms, although based upon sound scientific principles, must slavishly adhere to local values of inherent spectro-optical properties (which vary in time and space). However, a proper combination of field, laboratory, and modeling activity, coupled with a certain amount of *a priori* knowledge, can usually provide these values for such spectro-optical properties. Certain environments can also disqualify themselves from certain algorithm usage. For example, the operative word in multivariate optimization is *multivariate* in the sense that the system is composed of several variables as well as the numerical range of each variable comprising several values. For multivariate analyses to be used as an inverse modeling tool, indigenous sought-after environmental parameters (for example, concentrations of *chl*, *sm*, and *DOC*) must display a suitable range of numerical values. Similarly, environmental conditions may place certain restrictions on use of neural networks or empirical/semiempirical regressions. However, other options might be found appropriate for such environmental conditions.

The local atmosphere, however, is an environmental variable that remains an unsolved problem, although the principles of atmospheric physics and atmospheric chemistry are sound and well understood. As has been discussed, the "color" of the intervening atmosphere is an obfuscation that must be removed from the top-of-the-atmosphere biospheric "color" recorded by the satellite. Hampering the successful removal of this atmospheric signal is the fact that almost every aquatic component (chlorophyll, suspended minerals, or dissolved organic carbon) strongly absorbs short-wavelength visible blue light. Simultaneously, radiation backscattered from the atmosphere is highest at these blue wavelengths. Thus, modeling removal of blue light from water color recorded at satellite altitudes has, on numerous occasions, resulted in the determination of negative values of water-leaving radiances. In turn, this has produced erroneous estimates of aquatic chlorophyll concentrations.

The inability to determine precisely the distribution profiles of atmospheric aerosols (suspended atmospheric matter and liquid particles that exist in myriad diverse forms and diverse shapes such as smoke, water, clouds, dust, ashes, pollen, spores, etc.) within the atmospheric column linking the satellite sensor and the targeted water body is responsible for failure of many attempts to sense remotely the quality of case 1 and case 2 waters. Many workers have addressed removal of this unwanted intervening atmospheric signal in order to overcome this obstacle. A partial list of these workers would include Fraser et al. (1992, 1997); Gordon et al. (1983); Wrigley et al. (1992); Guzzi et al. (1987); Kneizys et al. (1983); Gordon and Wang (1992, 1994); and Ruddick et al. (2000), among others.

Figure 2.2 and Figure 2.3 in Chapter Two illustrated the frustrations that could emerge from the use of two different standard atmospheric correction algorithms with the same bio–geo-optical water quality model to convert level 1 SeaWiFS top-of-atmosphere radiance data into level 2 chlorophyll and suspended sediment concentration products. A further example of the atmospheric challenge to water quality monitoring from space was presented in Bukata et al. (2003). MODIS data were used in tandem with a ship-deployable suite of above- and below-surface spectrometers (acronym WATERS) to monitor Lake Erie chlorophyll remotely. This afforded an opportunity to intercompare the directly measured WATERS water-leaving radiance spectra and level 2 MODIS water-leaving radiance spectra derived from the "raw" spectral radiance data using the NASA atmospheric model to remove the atmospheric signal. Figure 3.3 (taken from their Figure 3) illustrates this intercomparison for Lake Erie stations 23 and 439 obtained during a "clear day" MODIS overpass in May, 2001. Clearly,

- Poor agreement is seen between the directly measured and atmospherically corrected level 2 MODIS water-leaving radiance spectra at both stations.
- WATERS-measured water-leaving radiance spectra at the two stations are consistent in spectral shape and spectral intensity.
- MODIS atmospherically corrected level 2 water-leaving radiance spectra of the two stations display consistency in neither spectral shape nor spectral intensity.
- An undercorrection for atmospheric intervention is suggested for the May overpass (although overcorrection was suggested for a subsequent MODIS overpass).

In situations in which such underflight aquatic data are taken, it might be appropriate to use the directly measured water leaving radiance spectra to convert MODIS (or any other satellite) top-of-atmosphere spectral data empirically into an atmospherically corrected water-leaving radiance spectra from which a multivariate bio–geo-optical model and nonlinear optimization

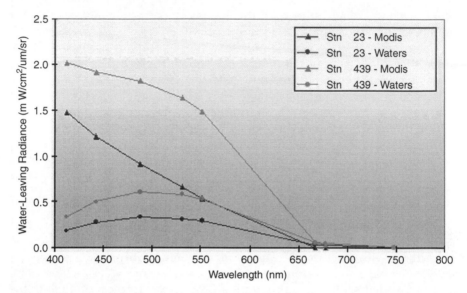

Figure 3.3 (See color insert following page 134.) Comparison of WATERS-derived water-leaving radiance spectra with those obtained from MODIS during May, 2001. (From Bukata et al., *Backscatter*, winter issue, 29–31, 2003.)

could generate chlorophyll concentration maps. Such an empirical conversion was performed on the MODIS May, 2001, overpass data. Figure 3.4, taken from Figure 4 of Bukata et al. (2003), illustrates the result of this conversion. The MODIS imagery was empirically modified using per-band regressions of level 1 MODIS radiances with WATERS water-leaving radiances. In this manner, the MODIS image was converted into an "equivalent WATERS space image."

Inverse optical modeling is computer intensive. Therefore, an unsupervised classification was performed on the equivalent WATERS space image and pixels were grouped into ten spectrally similar classes. Results of this simple unsupervised classification of Lake Erie pixels (based on ten water-leaving radiance spectra and use of underflight optical data to assist in removal of local atmospheric obscuring of true water color) are shown in the estimated chlorophyll concentration map of Figure 3.4. Obviously, full-pixel and full-spectral analyses would produce chlorophyll (and suspended mineral and dissolved organic carbon) concentration maps that would be more meaningful than Figure 3.4. However, cautious use of alternate approaches to reducing the impact of local atmospheric intervention can produce water quality products satisfying the advocated science/use of science compromise.

Atmospheric intervention also is understandably a major problem for terrestrial remote sensing. However, as discussed in Chapter Two, the atmosphere is especially problematical to aquatic remote sensing because the unwanted signal from the intervening atmosphere comprises upwards of

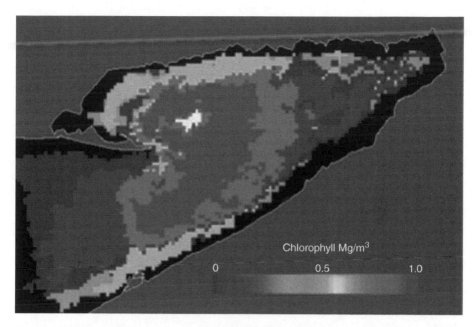

Figure 3.4 (See color insert following page 134.) Chlorophyll concentrations in Lake Erie as determined from a simple unsupervised classification of an "equivalent WATERS" image generated from level 1 MODIS data over Lake Erie. (From Bukata et al., *Backscatter,* winter issue, 29–31, 2003.)

~85% of the signal recorded at the satellite sensor. That is, at least 85% of the actual remotely sensed "color" is attributable to the atmosphere. Thus, an error of, say, 5% in estimating the impact of atmospheric aerosols can be approximately ten times as devastating to aquatic color as to terrestrial color.

The satellite must view one attenuating medium (natural water bodies) through another attenuating medium (the atmosphere) and not only distinguish one from the other but remove one from the other. Both attenuating media have their own sets of bulk and inherent optical properties. The science of photon propagation discussed in this chapter is also apropos to atmospheric absorption and attenuation of *extraterrestrial radiation* (total incident electromagnetic irradiance at the top of the atmosphere) and radiation emitted by or reflected from the Earth. Attenuating parameters include absorption from atmospheric gases (ozone, water vapor, and oxygen), scattering by air molecules (Rayleigh scattering that decreases with increasing wavelength), and absorption and scattering from aerosols.

For optically binary case 1 waters, some measure of success in removing the obfuscating effects of atmospheric "color" from the upwelling signal from ocean water has been achieved by the so-called "dark pixel" approach. Here, the entire signal from the atmosphere is attributable to one wavelength or one wavelength band. However, as focus has shifted towards continental waters, this approach has been systematically discredited — first, for selection of red wavelengths (Bukata et al., 1980) and, later, for selection of near-infrared wavelengths due to the measurable red and near-infrared radiance return from the chlorophyll and/or suspended mineral concentrations in inland and coastal waters. However, setting the aquatic return at wavelengths in the near-infrared region ~765 to 865 nm to zero (or some normalization ratio of two wavelengths) is still a component of atmospheric correction algorithms.

Reiterating, Rayleigh scattering, absorption by atmospheric gases and aerosols, and absorption by in-water chlorophyll and dissolved organic matter combine to reduce the water-leaving radiation in the short wavelength (blue) region of the visible spectrum. The atmospherically corrected water-leaving radiance as measured at the spacecraft very frequently becomes a negative radiance value when the following are incorporated into the analysis:

- The intractable patchiness of atmospheric aerosols
- The dark pixel normalization attempts
- The nearly minuscule intensities of water-leaving radiance spectra
- The paucity of indigenous CPA optical cross section values

To date, whatever successes have been obtained in atmospherically correcting data remotely acquired over mid-ocean waters have yet to be duplicated over the optically complex inland and coastal waters. The severity of this inability to do this over case 2 waters becomes heavily underscored by the realization that errors in the removal of the unwanted atmospheric signal from level 1 water quality products are cascaded throughout the conversion of level 1 water quality products to level 4 water quality products (see Section 1.6).

If there is a "take home message" in this section, it would simply be that the science of aquatic optics and that of atmospheric optics are alive and healthy. The perverse and recalcitrant natures of the aquatic and atmospheric environments, however, present stubbornly defiant challenges to this science. The challenges by the aquatic environment can be met, or at least satisfied, in most cases. The challenges by the atmospheric environment have as yet to be met satisfactorily. Circumventing or minimizing the impacts of these challenges must form part of the inevitable compromise between defendable science and defendable usage of science of Section 1.8.

Songwriter/singer Joni Mitchell, in her 1967 song "Both Sides Now," lamented how the illusionary nature of clouds presented an obstacle to things she might have otherwise done. The song could be an anthem for remote sensing scientists concerned with development of water quality products from satellite altitudes.

chapter four

Applications of water quality products to environmental monitoring

We have to go and save the Universe, you see…and if that sounds like a pretty lame excuse, then you may be right. Either way, we're off.

From *Life, the Universe, and Everything,*
Douglas Adams, 1982

In the Earth's environment, almost everything correlates with almost everything else. Furthermore, almost everything and almost everything else correlate with solar light. Extinguish light, extinguish biological life, at least as we now know it. The 1970s' attempts at environmental remote sensing became an incessant blur of regressing airborne and satellite apparent radiance data with almost everything and almost everything else, as long as each "everything" had been measured at the Earth's surface. Such a research plan immediately became eagerly implemented and the dreaded term "ground truth" was born, thereby plaguing, even now, the credibility of remotely sensed *estimates* of environmental parameters.

Of course, it was (or should have been) well known that such regression relationships are spatially and temporally dependent, reflecting as they do, the existing site-specific and interactive population concentrations and their biological status at a particular moment in time. This practice of plotting everything against recorded upwelling radiance has been more or less discarded in the remote sensing of case 2 waters. However, its surprising successes in case 1 waters (e.g., chlorophyll estimates based on regressing ratios of mathematically manipulated water-leaving radiances of differing wavelengths) would suggest that revisiting such simple approaches might possibly, under carefully considered occasions, be of consequence to the inevitable compromises we have been advocating between defendable science and defendable usage of that science.

However, revisiting must (1) conform to the view that case 1 waters and their optical behavior are the simple limits of case 2 waters and their optical behavior; and (2) be the direct antithesis to the fading but still prevalent view that case 2 waters can be considered "tweaking" limits of case 1 waters. The preceding possibilities should, therefore, be kept in mind throughout the upcoming discussions on compromises that must arise from the wants and needs of environmental end-users and the ability of remotely sensed inland and coastal water quality products to contribute to these wants and needs while retaining their scientific integrity.

4.1 What do environmental end-users want?

Aside: A possibly Machiavellian preamble

Asking the question, "What do environmental end-users want?" and then proceeding to answer it would certainly appear to be all or some of presumptuous, pompous, pedantic, pragmatic, insulting, irrelevant, and wrong. Certainly this appearance is accentuated when the question is first posed and the answer then given by the generators of environmental products promoted as pertinent to end-users' wants. Trying to deflect such demeaning adjectives (including self-serving) while simultaneously attempting to span the gap between science and applications is a challenge as well as the *raison d'être* of this chapter.

We do not have the credentials, the power, or the will to be perceived as lecturing decision- and policy-makers on what they should decide or legislate or on how they should carry out those mandates. We do admit, however, to a desire to influence them to consider space monitoring of the environment in general — and inland waters in particular — as an aid to executing their mandates. Thus, in an attempt to legitimize the question–answer format of this chapter we will

• Discuss environmental issues in broad generic terms, when that is possible.

- Utilize stated priorities of mandated environmental stewards as clues to what environmental end-users want and/or need.
- Utilize remote sensing science to suggest what information environmental end-users might be given.

Perhaps, in this manner, we might successfully sidestep the preceding adjectives (with the possible exceptions of "self- serving" and "pragmatic").

Environmental behavior is dictated by myriad physical and biogeochemical forcing functions acting upon myriad physical, chemical, and biological entities in myriad inter-related, quasi-independent, cumulative, and synergistic fashions, resulting in myriad time-series of physical and biogeochemical change. The fraction of environmental end-users who are scientifically focused are driven by the vast variety of cause–effect inter-relationships forming cyclical feedback loops responsible for these time-series of environmental change. Within this multifocused scientific community reside a component cognizant of aquatic optics and the capabilities and limitations of remotely sensing inland and coastal waters; a component cognizant of aspects of environmental behavior but with little interest in or knowledge of remote sensing; and a small component cognizant of aquatic optics, remote sensing, and uni- or multidisciplinary environmental science. Collectively, these scientific end-users

- Develop the science of remote sensing
- Generate and improve water quality products emergent from remote sensing of inland and coastal waters
- Apply these products to specific environmental issues
- Have already convinced themselves of the importance of monitoring case 2 water quality

Clearly, this group of end-users wants time-series of reliable and financially accessible data, accurate water quality algorithms and methodologies, and access to sophisticated data storage and data analyses facilities.

Furthermore, this scientific group of environmental end-users must present compelling evidence that its efforts are of consequence to a second group of environmental end-users. Comprising this second group of environmental end-users are the environmental managers/decision-makers and the political policy-makers who, in the majority of instances, are not cognizant of aquatic optics, remote sensing, or disciplinary and/or multidisciplinary environmental science. Also within this group are the executors of the decisions and/or policies that have already been made or have yet to be made. These end-users are generally politicians elected or appointed on a limited-term basis or public service or private sector employees hired within permanent or quasipermanent time-frames. They are charged with specific job responsibilities that usually require adherence to and continuation of established protocols and generally conduct their business without involvement of or thought of remote sensing as a tool essential to performance of their duties.

It is probably safe to assume that environmental decision-makers and policy-makers want answers to some variations of the questions, "How at risk is the local, regional, national, continental, and/or global environment?" "What will be the fiscal and cultural consequences of ignoring this risk?" "How may this risk be abated?" "What will be the fiscal and cultural consequences of implementing this abatement?" Although these questions are straightforward, their answers are invariably not. As already discussed, this is due to the frustrating vagaries of the aquatic and atmospheric compositions and optical properties, as well as the multidisciplinary cause–effect relationships governing status and evolution of these environmental sectors.

Dispassionate scientific searches for the answers must contend with the accepted tenet of multiple interactive environmental stressors (e.g., acidification, eutrophication, introduction of exotic species, fluctuations in climate parameters, solar cycle dynamics), which recognizes that the resultant response of an aquatic ecosystem to such conjoint stressors is a complex function of the individual, collective, and synergistic responses of the ecosystem membership. The term "environment" itself invokes a variety of passions and invective within the general populace, which can impede scientific quests for dispassionate answers to questions posed by environmental decision-makers and policy-makers. Furthermore, the term "risk" is ill understood and carries additional and possibly even greater passions and invective.

As a more specific illustration of what environmental decision- and policy-makers might want, we will refer to the list of ten priorities comprising the Environmental Scan for the Nature Research Agenda of Environment Canada (Environment Canada, 1999a, b). Although these were compiled with a direct focus on the Canadian environment, they would not be atypical of environmental priorities (other than perhaps hierarchically atypical due to highly unique site-specific environmental vulnerabilities) of other countries. Therefore, we feel that no major loss of generality would ensue from the use of these priorities as a template for what environmental decision- and policy-makers might want. These ten priority issues are:

- Wildlife/biodiversity
- Aquatic ecosystems (freshwater and marine ecosystems)
- Impacts of atmospheric change
- Impacts of land use
- Impacts of highly populated areas
- Impacts of pesticides and toxic substances
- Impacts of resource exploitation
- Impacts of exotic species
- Impacts of biotechnology
- Cumulative impacts

Despite the possible exclusion of topics pertinent to unique site-specific issues, the preceding priorities should provide justification for the use of remote sensing as a tool for ecological monitoring and, in particular, for the remote sensing of inland and coastal water quality as a component of this tool. Specific inclusion of aquatic ecosystems (freshwater and marine ecosystems) and cumulative impacts presents opportunities for dialogue with environmental decision- and policy-makers regarding case 2 water quality products.

A logical and realizable goal of a coordinated remote sensing program (aquatic, terrestrial, and wetland) would be the utilization of remote sensing sensors, vehicles, and methodologies to provide, on a continuing basis, a form of "state-of-the-environment" reporting mechanism based upon ecosystem change. Such a reporting mechanism, compatibly incorporated within existing ground-based monitoring-network reporting mechanisms, could provide valuable assessments of local, regional, national, or possibly global environments, depending upon the ability to cope successfully with geographic scaling factors.

A wide variety of remote sensing devices exists that is responsive to a wide variety of electromagnetic wavelengths in a wide variety of Earth-orbiting vehicles with a wide variety of target replication periods. The focus of this book is, essentially, upon the visible, near-infrared, and thermal infrared wavelengths and their roles in inland and coastal water quality monitoring. Although this focus will continue, the following points must be recognized:

- Water *quality* should not be considered in isolation from water *quantity.*
- As an integral component of the environment, water should not be considered in isolation from the adjacent or encapsulating terrestrial matter that can modulate and/or dictate its quality.
- Radar wavelengths play major roles in the remote monitoring of water quantity.
- Environmental health stressors are a combination of natural and societal activities that manifest as atmospheric forcing functions.
- The quality of inland and coastal waters is both a cause and an effect parameter that is a complex function of these atmospheric stressors as well as the spatially and temporally varying status of the terrestrial and wetland components of the local to global environment.

We will not even presume to solve these synergistic inter-relationships here. However, while pursuing the inevitable compromise that we have been advocating between defendable science and defendable usage of that science, we will stress the need to bring remote sensing of inland and coastal water quality — a component invariably absent from existing monitoring protocols — beneath the overarching umbrella of environmental monitoring. To do so, we shall first consider the ten priorities of the Environment Canada nature research agenda on an expanded-focus basis to include possible roles that remote monitoring in general (aquatic, terrestrial, wetland) could play in assisting the needs of these environmental priorities. Subsequently, we shall once again restrict our focus to the remote monitoring of inland and coastal water quality.

4.2 What do environmental end-users need?

Clearly, decision-makers and policy-makers concerned with fiscal and cultural consequences of environmental risk must regard human health and welfare as well as survival and sustainability of flora and fauna as their most important issues and concerns (and thus their *prima facie* needs). When they are in conflict, factors perceived to be detrimental to health and mortality must take precedence over those perceived to be detrimental to comfort and lifestyle; however, both sets must be considered. Such factors, however, are not always perceived dispassionately — with or without scientific input.

It is generally conceded that every element and/or chemical compound can be deleterious to human and ecosystem health if present in harmful to lethal dosages. What constitutes harmful to lethal dosages (with or without direct scientific input) is not generally conceded, however. Special-interest invective abounds in controversies arising from choices that must be made between perceived threats to health and welfare and perceived threats to comfort, lifestyle, and individual freedoms. Examples of these choices include (among many others):

- Second-hand smoke vs. smokers' rights
- Curtailing use of fossil fuel vs. increase in unemployment
- Removal of chlorofluorocarbons vs. reduction in standard of living
- Prohibiting use of soil-conditioning and crop-yield enhancement chemicals vs. financial insolvency of farmers
- Spraying mosquito populations vs. risk of pulmonary disorders
- Implementing environmental protocols vs. increased taxation

Science, unfortunately, is lodged somewhere in the midst of almost every human health and ecosystem health controversy. Even more unfortunately, it is invariably declared onside or irrelevant by one side or the other. Science is one of the many pawn weapons blatantly used by mind-sets on either side of an issue perception. (Principal weapons are emotion, zeal, vitriolic vituperation, and an unswerving belief in being correct and virtuous.) Certainly, hard science can sometimes readily validate the perception of threat by some special-interest groups. Sometimes, hard science can readily invalidate the perception of threat by some special-interest groups. However, definitive hard science can neither confirm nor deny many perceptions of threat. This situation arises from the vast spectrum of responses by humans, animals, and ecosystems to the controversial issues and a lack of sufficient all-encompassing observations to provide statistically reliable conclusions.

Not unexpectedly, therefore, the use and disposal of such substances as fuels, fertilizers, aerosols, paints, insect repellents, pharmaceuticals, batteries, and potential contaminants in general figure prominently within the "needs" of environmental decision- and policy-makers.

- *Presence, location, identity, concentration, impact, containment, and transport of toxic contaminants within aquatic, terrestrial, wetland, and atmospheric ecosystems*

Understandably, apart from the obvious inclusion of chemicals generally acknowledged to be toxic, a catalogue of environmental end-users' wants, at any time, emerges from a series of quasisubjective judgments predicated upon a series of ill-defined questions regarding well-defined objectives. Similarly, the corresponding catalogue of environmental end-users' needs, at that same time, would emerge from yet another series of quasisubjective judgments predicated upon those that have produced the catalogue of end-users' wants. Within this quasisubjective attempt at objectivity, we shall now briefly consider each Environment Canada priority from the perspective of using remote sensing as an appropriate tool for ecological monitoring.

4.2.1 Wildlife/biodiversity

Inherent to plant and animal biodiversity (species/populations) are the conjoint parameters that determine the ability of a continent's or an ocean's ecozones to sustain particular species (or sets of species) of plants and/or animals. *Ecozones* may be simply defined as ecologically distinct areas of the Earth's biosphere that have assumed permanently recognizable features due to synergies arising from interplay among climatic, geologic, soil, water, landform, vegetation, wildlife, and human factors (e.g., forest, tundra, wetland, mountain, grassland, desert). Changes in ecozones, therefore, manifest as changes in plant and animal diversities. Such changes might be consequences of undisturbed natural periods of environmental parameters. They might also be consequences of (1) anthropogenic disruptions to the natural periods; (2) unpredictable potentially catastrophic step-function climatic events; or (3) introduction of exotic species to an established dynamic equilibrium governing the extant biodiversity within the ecozone.

One observable parameter indicative of ecozone change is the bioproductivity of local vegetation (forests, natural grasslands, cultivated croplands, etc.), which, in turn, directly reflects the quantity and quality of indigenous quasipermanent and transient water. (*Bioproductivity* represents the rate at which energy is stored in an ecosystem or part thereof within a specified period of time. Energy is stored through the light-induced biochemical process known as *photosynthesis* discussed in Section 5.4. That section also shows that bioproductivity is comparable to the term *primary productivity* and that *primary production* (the total chemical energy in an ecosystem or part thereof at a specified instant of time) would be comparable to the term *bioproduction*.) Monitoring the temporal and spatial changes in bioproductivity or bioproduction of natural and/or cultivated vegetative cover, freshwater bodies, saline coastal waters, and wetlands could therefore provide insight into plant–soil–nutrient dynamics, animal populations, and vulnerability of wildlife habitat.

In such a manner, remote monitoring of terrestrial, aquatic, and wetland bioproductivity might be used to develop and routinely report a biodiversity index for the network of continental and coastal ecozones. These indices would, hopefully, be scientifically sound and readily understandable to an informed public. They could be consistent with such already existing environmental indices as relative humidity, wind chill, humidex, air quality index, UV-B index, and other indices routinely used on radio reports and television weather channels.

- *Biodiversity indices for land, water, and wetland*

A combination of interpretive techniques utilizing satellite data either/ both as a source of photographic imagery (photogrammetry) or/and digital data from which environmental parameter information must be inferred and extracted from models or algorithms can result in time-series classifications of ecozone type, composition, and geophysical features. Such classifications can provide valuable records of the changing face of continental landscape (e.g., tundra, rangeland, forest, cryosphere, rock formation, ground-level topography, shoreline evolution). However, perhaps even more important than delineating and monitoring changes in terrestrial and aquatic ecozones are delineating and monitoring changes in terrestrial and aquatic ecotones. An *ecotone* is the transition area between adjacent ecozones (e.g., between forest and wetland; wetland and standing water; or tundra and forest) that displays characteristics of each ecozone (including plant and animal species). Time-series of changing locations and changing geophysical and biophysical features of ecotones could play instrumental roles in climate impact assessment and suitability of wildlife habitat.

- *Changes in ecozone geophysical and biophysical features*
- *Changes in ecotone geophysical and biophysical features*
- *Landscape fragmentation and changes in wildlife habitats*

On regional to global scales, normalized difference vegetation index (NDVI) values have been commonly used to estimate temporal and spatial changes in relative terrestrial bioproductivity. The NDVI index is a simple ratio involving near-infrared radiances and visible radiances recorded at satellite altitudes; it is based on the well-known near-infrared wavelength reflectance from and visible wavelength absorption by chlorophyll-laden verdure. Although the generally used NDVI values are generated from the advanced very high resolution radiometer (AVHRR) on the series of Polar Orbiting Environmental satellites (POES), the 1-km footprint of the AVHRR stymies its use in many regional studies. Other satellites, including the Landsat series, are also utilized in such terrestrial bioproductivity estimates.

Almost all satellite-determined NDVI values employ the simple difference formula NDVI equals (near-infrared radiation minus visible radiation) divided by (near-infrared radiation plus visible radiation). Thus, NDVI values for a given pixel range from –1 to +1. However, no green leaves would yield a negative or even a zero NDVI value. Therefore, zero is generally considered as the absence of vegetation and index values approaching +1 are considered as indicative of high density of chlorophyll-rich leaves (i.e., high terrestrial bioproductivity). Despite the obvious scientific shortcomings of such a simple estimate of terrestrial bioproductivity, however, cautious use of satellite determinations of NDVI can benefit the needs of environmental decision-makers and political policy-makers and therefore satisfy the compromise between science and its usage. The addition of corresponding estimates of aquatic bioproductivity to these terrestrial NDVI values should form a good basis for monitoring the status of plant diversity and wildlife habitats. Insight into appropriateness of wildlife habitat to endangered species might then become an achievable goal.

- *Evaluation of potential habitats for endangered species*

Assume that a seamless linkage of bioproductivity estimates for land, water, and wetland may be established (an unconquered challenge, although it is a work in progress) and can routinely be provided for the totality of continental ecozones and/or ecotones. Then, this capability (along with a biodiversity index) could help develop long-term biodiversity strategies and implementation plans on local to global scales.

- *Provision of a component integral to the development of a long-term biodiversity strategy and implementation plan*

4.2.2 Aquatic ecosystems (freshwater and marine ecosystems)

The atmosphere, wildlife, and aquatic resources of a region, country, or group of countries comprise the mandates of environmental decision- and policy-makers. Inarguably, water quantity and water quality comprise principal limiting factors of ecosystem health. Thus, for remote sensing to function effectively as a tool for ecological monitoring, a sharp focus on the aquatic resources of the region/country becomes mandatory. Remotely sensed data (in pictorial and/or digital formats) acquired over a range of temporal and spatial scales can readily provide essential information on the physical aspects of water quantity. The tracking of onset, evolution, and recession stages of lake and river floods and flood-plains is now an operational reality — as is the tracking of onset, evolution, and recession of drought. Remote water level determinations — particularly in Arctic and Antarctic regions inaccessible to conventional water level survey techniques — can provide valuable input into water balance models to help predict and anticipate future water level fluctuations.

- *Inland and coastal water quantity (surface area, water level, water flow diversion, floods and droughts)*

Aircraft and satellite remote sensing have played and will continue to play an important role in hydrologic modeling, as demonstrated in several national and multinational programs (among others):

- Boreal Ecosystem–Atmosphere Study (BOREAS)
- International Geosphere–Biosphere Program (IGBP)
- Global Energy and Water Cycle Experiment (GEWEX)
- Land–Ocean Interactions in the Coastal Zone (LOICZ)
- World Hydrological Cycle Observing System (WHYCOS)
- Global Ocean Observations Study (GOOS)
- NASA's Earth Science Applications Research Program (ESARP)

Cleverly combining optical remote sensing (visible, infrared) with "all-weather," cloud-penetrating, active microwave remote sensing (e.g., radar) enables the delineation of water boundaries beneath dense canopies of vegetative cover. Analyses of historical satellite imagery delineating changes in water levels, water extent, and water boundaries benefit application and validation of current models of hydrological processes.

- *Hydrologic modeling*

Integral to the hydrologic cycle (and consequently to hydrologic modeling) is subsurface water (groundwater and soil moisture). Remote sensing at microwave wavelengths has been used to estimate soil water content, as well as snow water equivalents and cryospheric properties. In addition, visible and near-infrared wavelength bands of satellites such as Landsat have successfully used the vigor of regional natural and/or cultivated vegetation (à la NDVI-type analyses) to delineate the locations of discharge, recharge, and transition areas within freshwater basins. Knowledge of these subsurface water regimes provides the locations and directions of groundwater flow within those basins.

- *Delineation of regional groundwater discharge, recharge, and transition areas and their temporal changes*

Photogrammetry techniques applied to repetitive images on a variety of spatial and temporal scales provide evidence of changes in shoreline and coastal features.

- *Natural and anthropogenic changes in shoreline and coastal features*

Radarsat has become an effective operational tool for detecting and monitoring the location, progress, and impact of marine oil spills.

- *Monitoring for and of marine oil spills*

The POES/AVHRR and the SeaStar/SeaWiFS satellite sensor systems have been instrumental in detecting and monitoring the transport and evolutionary nature of large blooms of deleterious oceanic algae referred to as "red tides" in the offshore waters of the Pacific and Atlantic Oceans. Substantial blooms of deleterious blue-green algae have also been recently seen in the Laurentian Great Lakes.

- *Detection and spatial extent of deleterious oceanic and inland water algal blooms*

As discussed in Chapter Two, remotely sensed data acquired over a range of temporal and spatial scales can provide reliable water quality products for the optically complex inland and coastal waters. Knowledge of the inherent spectro-optical properties (optical cross section spectra) of indigenous organic and inorganic aquatic matter along with suitable bio–geo-optical models is essential.

- *Lake and coastal water quality (presence of algae, turbidity, dissolved organic matter)*

Knowledge of the changing values of near-surface chlorophyll concentrations enables estimates of the bioproductivity of standing water, just as knowledge of the changing values of near-surface dissolved organic carbon concentrations illustrates the roles of standing water in ecosystem carbon sequestration and/or release.

- *Bioproductivity of standing water*

We shall return to a more detailed discussion of the role of remotely sensed water quality shortly.

4.2.3   *Impacts of atmospheric change*

The physical and some of the chemical aspects of atmospheric change and their impacts on the climate/weather patterns and geophysical consequences of these impacts to continental and oceanic climes are topics that continuously affect day-to-day human activities. These topics traditionally hold public interest, and weather-related satellite products possess a distinctly marketable value. Thus, weather forecasting, global climatology, global change monitoring and modeling, air quality, and stratospheric chemistry are issues that are very well-served by Earth-orbiting satellite sensors responsive to a wide variety of electromagnetic wavelengths. A partial list of activities that pertain to direct monitoring of atmospheric parameters and phenomena or various aspects of global or regional weather change — and, by extrapolation, global or regional climate change — would include, but not be restricted to:

- *Assimilation of satellite data into weather forecasting models*
- *Assimilation of satellite data into climate and climate change models*
- *Assimilation of satellite data into ocean wave modeling and coastal and inland wave energy predictions*
- *Determination of extent and water equivalent of snow cover*
- *Monitoring and understanding the dynamics of variations within the Earth's cryosphere*
- *Tracking and forecasting the dynamics of sea ice*
- *Monitoring stratospheric ozone (in concert with ground-level monitoring of downwelling ultraviolet radiation)*
- *Tracking ongoing catastrophic climatic events (hurricanes, floods, forest fires, volcanic eruptions, etc.)*
- *Tracking ongoing catastrophic anthropogenic events (oil spills, tire fires, spatially extensive plumes of atmospheric toxic smoke, etc.)*

Perhaps the most familiar (and by extension the most important) "weather maker" from the perspective of day-by-day public concern would be the continuous variability of the physical upper atmosphere *jet streams* — a concept and a reality almost wantonly lending itself to the successful marketing strategies of weather-reporting media. Jet streams are high altitude (~9 to ~18 km) discontinuous bands of fast flowing air generally moving from west to east over northern mid-latitudes. The polar-front jet streams are located above areas of strong temperature gradients and separate the cold polar air at high latitudes from the warmer air at lower latitudes. Jet streams are also located above near-equatorial temperature gradients that separate tropical latitude air from mid-latitude air.

The rapidly changing dynamics governing local atmospheric pressure result in dramatic changes in the north–south and east–west configurations of these jet streams and, from resulting invasions of warm air into colder regions and cold air into warmer regions, bring about significant changes in local weather on scales of hours to days. Programmed as repetitive time-sequences in movie-like fashion, time-series of satellite-generated, near-real-time "maps" of the locations and serpentine windings of jet streams on television weather channels present a valuable remote sensing weather product to an end-user community. This large community is made up of weather-dependent professionals with educational, medical, law enforcement, emergency protocol, fuel supply, and other essential service responsibilities; military and civilian jet aircraft personnel; and the general public with plans for short- or long-distance travel or perhaps merely an outdoor barbecue or relaxing picnic in the park. These groups have already accepted the need for and utility of this remotely generated product.

Similarly, perhaps the most familiar (and by extension the most important) "weather maker" from the perspective of year-by-year public concern would be the recurrent disruptions to the ocean–atmosphere system in the tropical Pacific that have been termed *El Niño southern oscillations (ENSOs).* (El Niño is Spanish for *little boy* or *Christ Child.*) Under non-El Niño conditions, trade winds blow westward from the Galapagos Islands across the tropical Pacific, piling up water in the west Pacific (ocean surface ~0.5 m higher and ~8°C warmer at Indonesia than at Ecuador). The cooler water around South America is the consequence of nutrient-rich cold water upwelling from deeper ocean depths. This water supports high primary productivity and diverse marine ecosystems and is consequently a major source of food-fish. The west–east Pacific water temperature gradient accounts for high rainfall in Indonesia and relatively little rainfall in equatorial South America.

During an El Niño, trade winds weaken in velocity in the central and western Pacific, resulting in a depression of the oceanic thermocline in the eastern Pacific and an elevation of the oceanic thermocline in the western Pacific. East–west oscillations are affected. Vertical migrations of oceanic isotherms during an El Niño can be quite substantial. The resultant reduction of upwelled nutrient-rich cold water around western South America can dramatically reduce the hitherto highly productive coastal waters, thereby disrupting energy-transfer mechanisms (nutrient driven and predator driven) throughout the entire marine food chain. Also, the warmer surface waters around South America result in increased rainfall. The rainfall pattern established during non-El Niño years (increased rainfall as the equatorial Pacific is traversed east to west) is disrupted and basically reversed (decreased rainfall as the equatorial Pacific is traversed east to west). Flooding has occurred in Ecuador and Peru during an El Niño. Droughts have occurred in Indonesia and Australia during an El Niño.

These eastward displacements of sources of Pacific Ocean heat to the atmosphere result in major abnormalities to ocean–atmosphere dynamics during ENSOs that manifest large changes to atmospheric circulation patterns. These, in turn, can manifest a broad spectrum of changes to regional climate patterns in regions far removed from the equatorial Pacific. Currently, it is evident in Australia that research imperatives with a major focus on freshwater management can arise from such major weather makers and climate modulators. At the end of a recent prolonged El Niño event and the worst drought since European settlement in Australia, water management has become the single biggest political issue and a matter of intense public concern and debate.

Although ENSOs are recurrent, they are not periodic, recurring at irregular intervals ranging from 2 to 10 tears, with no two ENSOs alike. An El Niño event is considered a "warm" oscillation event due to resultant unusually warm ocean temperatures in the equatorial Pacific. An anti-El Niño event (known as La Niña, which is Spanish for *little girl*) is considered a "cold" oscillation event because it is characterized by unusually cold temperatures in the equatorial Pacific.

Despite the brevity of this discussion, it is clear that these large-scale trans-Pacific thermal oscillations that have an impact on global weather and climate are phenomena of great scientific and public interest. Thus, satellite ocean-thermal data, readily translatable into impressive animated time-series of oscillatory temperature patterns spanning a significant portion of the Earth's circumference, possess an end-usership convinced of the market value of its weather- and climate-focused products.

- *Scientific and public appeal of large scale atmospheric change (jet streams) and large scale ocean temperature change (ENSOs)*

Activities and products dealing with dramatic large-scale phenomena that can be readily transformed into entertaining media-friendly formats and presented in an impressive and easily understood manner can usually attract the attention of a well-defined community of end-users and clients. Remote sensing of inland and coastal water quality has yet to find its equivalent to a jet stream or an ENSO. However, if compelling computer-generated animations of sediment transport or water color change could be presented (at least to decision- and policy-makers) in a manner convincingly linked to public worth and health, value-added remote sensing of environmental ecosystems might become a salable reality.

Aside: A possibly (but not intentionally) impolite sidebar

The marketing strategies of the weather forecasting media are excellent examples of how compromises between defendable science and defendable usage of science and what end-users want and what they can be given have been successfully implemented to add public value to remotely derived products. Although weather forecasting into the very immediate future has displayed far from enviable accuracy and has many times provided humorous material for stand-up comic routines, it continues to enjoy public trust because it markets a product whose shortcomings are readily forgiven by the same public that it inadvertently deceives.

However, as we have discussed, weather forecasting is promoted as products "sold" in attractive, eye-catching, entertaining packaging. Obvious inaccuracies that are easy to document (diurnal local temperature swings of tens of degrees Celsius; local atmospheric disturbances and dynamic interactive weather systems being responsible for predicted storm warnings manifesting as UV-B alerts; predicted extensive snow manifesting as local rain showers; etc.) do not appear detrimental to public acceptance of atmospheric change products predicting short-term weather change. This acceptance is undoubtedly aided by the relative ease with which these erroneous predictions may be explained and excused by directly recorded, readily understandable, real-time observations aided by viewer-friendly serpentine meanderings of jet streams across television screens.

However, public acceptance of atmospheric change products predicting long-term climate change has not been as readily forthcoming. Perhaps this is because long-term climate inaccuracies are not obvious and easy to document. Perhaps this is because day-to-day temperature swings of tens of degrees Celsius coupled with predictive temperature errors of ±5°C or more in local temperatures belie belief in predicted global temperature changes of ~1°C in 20 years. Perhaps this is because climate change models generally relying upon "what if" scenarios do not offer defendable science and their outputs, as suggested earlier in Section 4.2, become weapons in climate change (or its improperly considered alter-ego "global warming") controversies. Perhaps this is because anthropogenic climate change (global warming) has fallen into the confounding domains of social, cultural, environmental, and political activism and rhetoric. Perhaps this is because of all of these reasons and more.

Nonetheless, lessons are to be learned here. To advance remotely acquired water quality products, the short-term marketing strategies (based on a product that the public is willing to accept despite its "obvious yet forgivable" shortcomings) that have proven to be successful for promoting atmospheric change products should be adopted for promoting water quality change products, if possible. However, comparable doubts might well be cast upon the validity of long-term predictions of water quality change. Perhaps this suggests that, at least for the present, the best chance for public acceptance of space observations of inland and coastal water quality would arise from concentrating on archival and recent water color data to illustrate how (or at least *that*) we got from "before" to "now" rather than trying to predict where we are going from "now" to "later."

Furthermore, local water quality is linked not only to local weather, but also to local use/abuse resulting from human demands on the watershed. Therefore, short-term water quality issues of public concern should be seriously addressed (eutrophication, remediation, and impact assessment of proposed agricultural, industrial, recreational, cultural, or residential programs). This is preferable to attempting to invoke local water quality products as a long-term climate change predictor. Admittedly, this is debatable.

Impacts of atmospheric change are one of the principal causes of ecosystem change on local to global scales. By simultaneously providing "snapshots" of atmospheric and ecosystem change, remote sensing has the potential to provide insight into cause–effect relationships dictating the state of the regional environment. Remote sensing has the ability to provide small- to large-area "snapshots" of the changing totality of biospheric parameters interplaying beneath the remote sensor. Therefore, it also provides a spatial framework within which local and regional ground-acquired monitoring information can be displayed, evaluated, and extrapolated (the basis of geographic information systems, GIS, analyses that merge point source ground vector data with remotely acquired raster data). Remote sensing capabilities can thus be used in the absence of concurrent ground information or can be extended when used in concert with concurrent ground information.

Changes in the spectro-optical properties of the fabric comprising continental and aquatic ecozones and/or ecotones are indicative of the consequences of changing cause–effect relationships governing the ecosystems sustained there. Thus, for remote sensing to be applicable to monitoring the ecological impacts of atmospheric change, the operational and developmental remote sensing activities of weather and climate atmospheric monitoring (listed at the beginning of this section) must interact with suitable remote sensing contributions to the biodiversity/wildlife and aquatic ecosystems priorities listed in Section 4.1. That is, multispectral data from a variety of passive and active satellite sensors must be used to classify and evaluate the terrain for:

- *Changes in ecozone/ecotone location and dimension*
- *Changes in water quantity, discharge, and distribution*
- *Changes in timing of ice formation and breakup*
- *Changes in terrestrial, aquatic, and wetland bioproductivity*

This is in addition to the development of climate change and impact indices (in a manner similar to and consistent with the suggested biodiversity index) based upon ecosystem response.

- *Development of climate change indices for terrestrial, aquatic, and wetland ecosystems*

4.2.4   Impacts of land use

The maximum benefit of utilizing remote sensing in its photographic form has traditionally emerged here. When used in conjunction with time-series data collected over the past few to 30 years of environmental satellite sensing, photographic imagery can be used to provide insight into the environmental dynamics responding to urban growth over the past quarter century. Historical airborne imagery and site-specific, ground-acquired data sets, when merged with satellite data (through GIS methodologies), can often "hindcast" such insight further back in time.

- *Dynamics of urban growth*

The Landsat Multispectral Scanner and Thematic Mapper, as well as other passive remote sensors, have provided agricultural managers and scientists with invaluable soil evaluation data. When used in conjunction with climatic and hydrologic models, these data can provide detailed analyses of erosion risk.

- *Erosion risk modeling in agricultural lands*

In a similar manner, remote sensing has been a valuable tool for national land inventory on a global scale.

- *National inventory of landscape classification and change*

A seamless linkage of terrestrial, aquatic, and wetland bioproductivity, coupled with the temporal changes therein, can contribute to impact assessments of existing and/or proposed land use activities (i.e., initial environmental conditions and/or resulting environmental conditions).

- *Change in bioproductivity of terrestrial (forest, farms), aquatic, and habitat resources adjacent to specific existing or proposed land use activities (i.e., impact assessments)*

Spatial extents of the impacts of land use activities can also be readily obtained from a seamless linkage of environmental bioproductivity.

- *Changes in areal extent and vigor of natural resources*

With the ability to assess environmental bioproductivity from a time-series of environmental overviews, satellite data may provide an effective means of assessing the successes or failures of designated environmental rehabilitation programs (e.g., toxic waste disposal and clean-up, reforestation, habitat reclamation, coastal shoreline regeneration, etc.).

- *Success/failure of designated environmental site-rehabilitation programs*

4.2.5   Impacts of highly populated areas

A major impact of highly populated areas (apart from the initial displacement of a natural wildlife ecosystem by an urban/suburban ecosystem) is the evolution of "haze hood" dynamics; this is generally thermal and chemical in nature. Several satellites provide very accurate surface/near-surface temperature values over a variety of temporal and spatial scales. Chemical dynamics, however, must be inferred through measurements of the physical properties of the atmosphere and spectral radiation from Earth's terrain. Atmospheric parameters of consequence are the density, distribution, and absorptive/reflective properties of aerosols. Environmental parameters of consequence are air and water quality, water quantity, and changing landscapes.

- *Temporal and/or spatial changes in thermal properties of the region and physical-chemical properties of atmospheric aerosols (smog, haze, etc.)*

Using the synoptic overview advantages of repetitive satellite coverage, a direct measure of the ongoing consequences of high population densities may emerge from the temporal and spatial changes in the spectro-optical contrast between the highly populated area and its environs.

- *Temporal and/or spatial changes in spectral contrast between the highly populated area and its surrounding environment*

4.2.6   Impacts of pesticides and toxic substances

Without *a priori* undisputed knowledge of concurrent directly measured concentrations of an unambiguously identified toxic substance, monitoring by remote sensing cannot identify pesticides and toxic substances. Nor can it provide scientifically defendable estimates of toxic contaminant concentrations once their presence and composition have been otherwise ascertained. It is, therefore, important to realize that:

- *Neither unambiguous identification nor concentration estimates of airborne or water-borne toxic substances is possible using remote sensing as a stand-alone monitor*

However, if concentrations of the effluents from known point-source injections of toxic matter into regions of the environment are of sufficient intensity and spectral signature to be observable at remote sensing altitudes (e.g., emissions of factory chimneys, extensive atmospheric plumes from industrial complexes or cities, mine-tailings in lakes and streams, colored effluents from processing plants, etc.), the propagation paths of these injections can be readily tracked and changes in their relative intensities might be cautiously inferred. Although the point of injection and aspects of its subsequent progress may be tracked by remote sensing platforms, the chemical composition and concentrations of these effluents cannot be determined from remote sensing. Atmospheric and aquatic suspended inorganic particulate matter serve as transport vehicles for most toxic chemical contaminants. Remote monitoring of sediment transport in lakes and rivers in some instances could, with appropriate *a priori* knowledge, make a vicarious second or third order contribution to toxic fate analyses.

- *Direct observations of atmospheric aerosols or possible point-source toxic aquatic inputs (atmospheric plumes and other "end-of-pipe" effluents)*
- *Indirect inferences of toxic fate through direct observations of atmospheric or aquatic inorganic particulate transport*

Impacts of pesticides and toxic substances will manifest as changes in the physical and biophysical properties of the terrestrial, aquatic, and/or wetland ecosystems. However, such changes are the result of conjoint and interactive environmental stressors. The challenge to deconvolve environmental stress into responsible stress agents remains.

- *Changes in terrestrial, aquatic, and wetland bioproductivity*

4.2.7   Impacts of resource exploitation

Resource exploitation (along with the economic, industrial, and cultural evolution associated with resource exploitation) is the major contributor to non-natural environmental stress. Ecosystem change and possible climate change on a local scale are interactive and synergistic with this environmental stress. Existing resource exploitation activities (mining, logging, fishing, hunting, water diversion, construction, inland or offshore oil drilling, etc.) are readily located and identified from satellite altitudes. As discussed, the greatest virtue of repetitive monitoring from satellite altitudes is its ability to supply site-specific time history of environmental change. Remote sensing can readily provide direct numerical values of physical and geometric properties of the biosphere being monitored. However, the greatest limitation of remote sensing is its inability to provide *direct numerical values of environmental parameters of the biosphere that are nonphysical.* We shall revisit this concept in the discussion of cumulative impacts in Section 4.2.10.

The capability of satellites to provide a site-specific time history of environmental change affords opportunities to use remote sensing as a tool for assessing the impacts of ongoing and proposed activities centered on the exploitation of natural resources. By assessing the Earth's biosphere on a range of spatial and temporal scales, changes — or, more importantly, changes in the *rate of change* — in the spectro-optical properties of ecosystems at the resource exploitation site and its environs are invaluable to impact assessment. Changes in land use and water use result in short-term and/or long-term changes in terrestrial, aquatic, and/or wetland ecosystems. Thus, transformations of these ecosystems are key indicators of environmental impact.

- *Transformation (enhancement or degradation) of terrestrial, aquatic, and/ or wetland ecosystems*

As is the case for all environmental stressors, monitoring of landscape features, bioproductivity, water quantity and quality, wildlife habitat, and climatic patterns provides information essential to the assessment of ecosystem impacts.

- *Temporal and/or spatial changes in environmental landscape*
- *Temporal and/or spatial changes in landscape bioproductivity*
- *Temporal and/or spatial changes in water quantity and quality*
- *Temporal and/or spatial changes in wildlife habitat*
- *Temporal and/or spatial changes in climatic patterns*

4.2.8   Impacts of exotic species

Invasions by exotic species of flora or fauna into an ecosystem that has developed a stationary or quasistationary equilibrium based upon adaptation to the lifestyle dynamics of its current membership result in disruptions to that equilibrium. Such disruptions are not anticipated, may be rapid or slow in developing, and result in new equilibria that may be considered good or bad. Recent examples of exotic species introduced into Canadian ecosystems include the zebra and quagga mussels and purple strife. The impacts of these and other exotic species are currently being evaluated from extensive field and laboratory studies. Data from site-dedicated aircraft and some satellite data are also being used.

Exotic species are environmental stressors that synergistically interact with the suite of conjoint stressors that combine to produce ecosystem change. Again, the main role of remote sensing in ecosystem impact assessment is monitoring change and rate of change in spectro-optical properties of those ecosystems. As in the cases of impact assessment from other environmental stressors, remote monitoring must focus upon parameters and functions that govern and measure ecosystem biodiversity and the stability of equilibria established there. Special attention should be given to the plight of endangered species.

- *Temporal and/or spatial changes in terrestrial, aquatic, and/or wetland bioproductivity*
- *Temporal and/or spatial changes in terrestrial, aquatic, and/or wetland habitat (suitability for endangered species)*
- *Temporal and/or spatial changes in water quantity and clarity*
- *Temporal and/or spatial changes in inland cover type and vigor*

4.2.9   Impacts of biotechnology

Alterations to genetic fabrics of naturally evolved flora and fauna induced by biological engineering are currently receiving sharp focus from biotechnological studies at research institutes in several countries. No biological parameter *per se* is measured from remote sensing platforms because such parameters must be inferred from radiometric measurements of spectral return from the environment; therefore, these inferred values of biological status represent values that cannot unambiguously be identified as pertinent to genetically altered or unaltered components of the monitored biosphere. Unless *a priori* knowledge exists of the biospheric target being monitored, remote sensing cannot ascertain whether or not its past history has been biotechnologically rewritten.

A good example of how remote sensing can be "fooled" by human intervention within an ecozone is illustrated in hydrology. For some time, it has been known that chlorophyll concentrations within vegetative canopies of freshwater basins can be successfully used to indicate the proximity of the water table to the Earth's surface (e.g., Bukata et al., 1978). However, in the absence of ancillary ground-acquired information, whether this proximity is the result of natural groundwater flow or installation of subsurface tiles to improve agricultural yield is moot. It is difficult to conceive of an operational role of satellites in the expanding field of biotechnology. If a role does exist for remote sensing, it might be a research role in controlled experiments involving large-area intercomparisons of the behavior of conventional and bioengineered genetically altered flora and fauna.

- *Other than possibly playing a research role in experiments involving controlled large-area intercomparisons of conventional and bioengineered genetically modified plants, animals, and/or organisms, remote sensing does not as yet possess operational applications.*

4.2.10   Cumulative impacts

As mentioned in nearly all of the preceding sections on environmental priorities, ecosystem stress is a consequence of a suite of environmental stressors brought about by some combination of changes in the composition and physical/chemical properties of the atmosphere; introduction of unexpected exotic species (including humans) into the existing membership of ecosystems; and recurring, nonperiodic step-function climate events of varying magnitude and location. Consequences of this ecosystem stress are reflected by spatial and temporal changes in the chemical, physical, and/or biological properties of the ecosystems as well as changes in the rates at which these changes occur.

When synoptically recorded at altitude, the electromagnetic radiation emanating from a monitored ecosystem comprises spatial "snapshots" of the integrated consequences of all the environmental stressors in play at the time at which the remote observation was performed. Thus, the major virtue of remote sensing (as we have continually reiterated) in monitoring environmental ecosystems is its ability to illustrate changes and rates of those changes within the cumulative impacts of environmental stressors on that ecosystem.

- *Remotely sensed data are spatial "snapshots" of the integrated consequences of all environmental stressors governing behavior of the ecosystem at a given moment in time.*
- *Time-series remotely sensed data are spatial "snapshots" of the changes occurring within those integrated consequences.*

As a very brief reprise of the discussions in Section 1.4, it is most ironic that the greatest virtue of remote sensing — its ability to provide time-series records of the integrated impacts of extant suites of environmental stressors — also provides its greatest challenges and/or liabilities: the needs to

- Financially access such time-series records
- Develop and apply biophysical, chemical, and geological models to extract estimates of nonphysical parameters from copious streams of spectro-optical data
- Develop and apply multivariate analyses techniques to deconvolve responses to specific stressors from a cumulative and synergistic response to a totality of environmental stressors
- Develop and apply criteria by which anthropogenic stress response may be separated from natural stress response

This irony is the *raison d'être* behind multidisciplinary and multinational research being conducted globally in government, university, and private sector facilities and directed towards applications of remote sensing to the monitoring and assessing the cumulative impacts of environmental stressors.

- *Remote sensing provides an opportunity for assessing cumulative impacts of a suite of environmental stressors. Such assessment may be integral to the separation of natural stressor impacts from anthropogenic stressor impacts*

4.2.11   A summary of what environmental end-users might need

Reiterating the emphasized points of Sections 4.2.1 to Section 4.2.10, as an indication of the collective needs of environmental end-users, we have the following compendium:

- Presence, location, identity, concentration, impact, containment, and transport of toxic contaminants within aquatic, terrestrial, wetland, and atmospheric ecosystems

- Biodiversity indices for land, water, and wetland
- Landscape fragmentation and changes in wildlife habitats
- Evaluation of potential habitats for endangered species
- Development and implementation of a long-term biodiversity strategy
- Inland and coastal water quantity (surface area, water level, water flow diversion, floods and droughts)
- Hydrologic modeling
- Delineation of regional groundwater discharge, recharge, and transition areas and temporal changes therein
- Natural and anthropogenic changes in shoreline/coastal features
- Monitoring for and of marine oil spills
- Detection and spatial extent of deleterious oceanic and inland water algal blooms
- Lake and coastal water quality (presence of algae, turbidity, dissolved organic matter)
- Assimilation of satellite data into weather forecasting models
- Assimilation of satellite data into climate and climate-change models
- Assimilation of satellite data into ocean-wave modeling and coastal and inland wave-energy predictions
- Determination of extent and water equivalent of snow cover
- Monitoring and understanding the dynamics of variations within the Earth's cryosphere
- Tracking and forecasting the dynamics of sea ice
- Monitoring stratospheric ozone (in concert with ground-level monitoring of downwelling ultraviolet radiation)
- Tracking ongoing catastrophic climatic events (hurricanes, floods, forest fires, volcanic eruptions, etc.)
- Tracking ongoing catastrophic anthropogenic events (oil spills, tire fires, spatially extensive plumes of atmospheric toxic smoke, etc.)
- Large scale atmospheric change (jet streams) and large scale ocean temperature change (ENSOs)
- Changes in ecozone geophysical and biophysical features
- Changes in ecotone geophysical and biophysical features
- Changes in water quantity, discharge, and distribution
- Changes in timing of ice formation and break-up
- Changes in terrestrial, aquatic, and wetland bioproductivity
- Development of climate change indices for terrestrial, aquatic, and wetland ecosystems
- Dynamics of urban growth
- Erosion risk modeling in agricultural lands
- National inventory of landscape classification and change
- Change in bioproductivity of terrestrial (forest, farms), aquatic, and habitat resources adjacent to specific existing or proposed land use activities (i.e., impact assessments)
- Changes in areal extents and vigor of natural resources

- Success or failure of designated environmental site-rehabilitation programs
- Temporal and/or spatial changes in thermal properties of the region
- Temporal and/or spatial changes in atmospheric aerosols (smog, haze, etc.)
- Temporal and/or spatial changes in spectral contrast between the highly populated area and its surrounding environment
- Unambiguous identification and reliable concentration estimates of airborne and water-borne toxic substances
- Direct observations of atmospheric aerosols or possible point-source toxic aquatic inputs (atmospheric plumes and other "end-of-pipe" effluents)
- Fate of toxic chemicals in the environment
- Transformation (enhancement or degradation) of terrestrial, aquatic, and/or wetland ecosystems
- Temporal and/or spatial changes in environmental landscape
- Temporal and/or spatial changes in landscape bioproductivity
- Temporal and/or spatial changes in water quantity and quality
- Temporal and/or spatial changes in wildlife habitat
- Temporal and/or spatial changes in climatic patterns
- Temporal and/or spatial changes in terrestrial, aquatic, and/or wetland bioproductivity
- Temporal and/or spatial changes in terrestrial, aquatic, and/or wetland habitat (suitability for endangered species)
- Temporal and/or spatial changes in water quantity and clarity
- Temporal and/or spatial changes in inland cover type and vigor
- Biological engineering-induced alterations to the genetic fabrics of naturally evolved flora and fauna
- Integrated consequences of the suite of environmental stressors governing behavior of a given ecosystem at a given moment in time
- Separation of natural environmental stressor impacts from anthropogenic environmental stressor impacts

As conceded at the start of this exercise, our assembled list of environmental end-users' needs result from quasisubjective attempts at objectivity. The preceding list may claim a certain degree of relevance to what would be required by end-users charged with environmental stewardship (that can coexist with lifestyle, health, economy, and societal customs). Obviously, however, we cannot claim that we have exhausted the pool of relevance. Indeed, other end-user needs would emerge as a site-specific issue was addressed or would appear unexpectedly. Nonetheless, we will consider this list in a manner consistent with the view that remote sensing is a valuable, although underutilized and underappreciated, tool for environmental monitoring.

4.3   What can environmental end-users be given?

Heading the list of environmental end-users' needs is their need to be aware of the location, identity, concentration, impact, containment, transport, and fate of toxic contaminants in aquatic, terrestrial, wetland, and atmospheric ecosystems. It is unfortunate that remote sensing of environmental color and temperature cannot contribute directly to this most important end-user need. The optical cross section spectra of toxic substances are insufficient to have a significant impact on aquatic or terrestrial color *per se,* even in large doses. Thus, toxic contaminants cannot be considered an aquatic component in inverse bio-optical modeling of environmental color. Therefore, remote sensing on its own cannot provide unambiguous information regarding the presence, location, identity, impact, or fate of toxic chemicals within the environment. However, because it may be readily assumed that the presence of toxic chemicals manifests as reduced environmental vigor, remotely monitoring environmental bioproductivity may possibly enable remote sensing to play a peripheral and subordinate role to ground-based chemical monitoring.

The principal focus of this book is on applications of products resulting from remote sensing of natural inland and coastal water quality. However, sight must not be lost of the facts that remote sensing provides time-series of environmental change and that these environmental changes represent interplay among aquatic, terrestrial, and wetland ecosystems. Although remote sensing is underutilized in ecosystem health assessment, it is nonetheless used in terrestrial, aquatic, and wetland assessment. A variety of nonaquatic, quasioperational monitoring activities (based on existing models and algorithms) exists, including but not necessarily restricted to:

- Terrestrial applications: these are well advanced and all enjoy quasi-operational status:
  - Spatial and temporal changes in size and locations of natural and cultivated land cover (e.g., forest, farmland, tundra, water, ice, desert)
  - Spatial and temporal changes in environmental ecozones
  - Supervised and unsupervised classification of forest species and agricultural crops
  - Estimates of vigor of forest, agricultural crops, and other ground verdure
  - Tracking progression and extent of landscape modifications associated with floods, forest fires, insect infestations, tectonic activity, hurricanes, and other destructive step-function phenomena
  - Geological mapping and mineral exploration
- Wetland applications: these are in relative infancy with only the first two enjoying quasioperational status. The remainder are works in progress:

- Spatial and temporal changes in size and locations of coastal and inland wetlands
- Supervised and unsupervised classification of wetland types
- Temporal and/or spatial changes in wildlife habitat
- Evaluation of potential habitats for endangered species
- Monitoring progress on rehabilitation of wetlands and wildlife habitat

Environmental remote sensing of inland waters is currently subordinate to environmental remote sensing of terrestrial features; this is evidenced by the fact that bioproductivity maps of national land masses continue to display inland waters as merely black holes in the terrestrial landscape. In order to attract influential end-users to water quality remote sensing products, it is essential that these black holes be convincingly filled.

Due to the escalating prominence of inland and coastal waters in the overall issues of carbon exchange and bioproductivity, remotely acquired inland and coastal water quality products must be seriously regarded outside the scientific community that is producing them. To achieve such recognition, changes in bioproductivity of inland and coastal waters must be integrated into impact assessments of regional climate change (and other environmental stressors) in the same manner as the bioproductivity of mid-ocean waters has been integrated into impact assessments of global climate change. Therefore, the following list of currently possible applications must be considered in context with the previous lists of terrestrial and wetland applications.

- Aquatic applications:
  - Mapping chlorophyll concentrations and primary productivity of mid-oceanic (case 1) waters (operational)
  - Mapping chlorophyll concentrations and primary productivity of the optically complex inland and coastal (case 2) waters (not yet operational)
  - Delineating the presence of algae and inorganic turbidity in inland waters
  - Monitoring the extent and progress of inland and marine oil spills
  - Monitoring the extent and progress of blue-green algae and "red tides" (large blooms of deleterious ocean and coastal algae)
  - Delineating regional groundwater discharge, recharge, and transition areas and their temporal changes
  - Monitoring inland and coastal water quantity (surface area, water level, water flow diversions, ice onset and break-up, sea-ice floe migration)
  - Recording natural and anthropogenic changes in shoreline/coastal features

Notice the comforting overlap within the preceding lists of what environmental decision-makers and policy-makers might want and what they might be given by environmental remote sensing. This overlap is rendered all the more comforting when the following abbreviated list of cross-cutting issues is considered:

- Cross-cutting applications (based upon the preceding three lists):
  - Detection of natural and anthropogenic changes in local weather and climate
  - Detection of environmental impacts resulting from local climate change
  - Fiscal, social, cultural, and political consequences of local environmental change
  - Early warning of impending accelerated environmental change
  - Proper environmental stewardship through development of sound protection, mitigation, and adaptation strategies and policies

4.4   Benefits of incorporating remote estimates of water quality into environmental monitoring protocols: value-added remote sensing

Compelling value-added remote sensing resulted in acceptance of atmospheric and ocean remote sensing products by environmental decision-makers and policy-makers concerned with weather, climate, agriculture, mining, forestry, and resource inventory. Compelling value-added remote sensing of inland and coastal waters must now be presented to these and other end-users of environmental information. Inarguably, the quality of inland and coastal waters is vital to the health of the human population and the environment of every country. It is also vital to countries whose strength and survival involves a resource-based economy. A tool to monitor changes in water quality, therefore, would be a great asset in protecting and sustaining this resource. Forewarned is forearmed; early warning of environmental stress or increased ecosystem vulnerability in a particular region could enable governments and other agencies to act before the situation worsens.

From Environment Canada (1991):

- Saltwater oceans and seas contain ~95.1% of the world's water supply.
- Freshwater lakes, rivers, and underground aquifers hold only ~3.5% of the world's water supply.
- Of the world's freshwater, ~66% is found underground.
- Of the world's freshwater, ~30% exists as ice in the form of glaciers and ice caps.

This would roughly suggest that, on a global scale, only 4% of 3.5%, or 0.14%, of world's water supply is readily available as standing water in lakes or running surface water in rivers. Also from Environment Canada (1991), the Laurentian Great Lakes straddling the Canada–U.S. boundary and Lake Baykal in Russia together contain 50% of the world's total fresh lake water. Thus, less than half of the 0.14% of the world's available standing and running freshwater is available to global regions outside North America and Russia.

From Pernetta and Milliman (1995), about 60% of the human population lives in the ocean coastal zone, defined as the region between 200 m above and 200 m below mean sea level, and this coastal zone comprises 8% of the ocean's surface. This 60% of the world's population resides in two-thirds of the world's major cities. Thus, 40% of the world's population and one-third of the world's major cities depend upon freshwater lakes and rivers as their sole supply of readily-available water. Clearly, large masses of global citizenry are dependent upon this recyclable, life-giving, pollutant-vulnerable, globally scarce, and nonhomogeneously distributed natural resource.

Natural patterns of surficial run-off and subsurface dispersion (including transport and dispersion of sediments and nutrients) are modified by human activities in aquatic catchments (forestry, land clearing, agriculture, mining, urban/suburban dynamics, industrial development). Coastal development along oceans, lakes, rivers, and streams leads to modifications (generally losses) of wildlife habitats, changes in aquatic flushing rates, and direct injections of nutrients and toxic contaminants into the circulation patterns of the near-shore and pelagic waters. Inland and coastal ocean waters contribute a disproportionate share (over 90%) of the world's food-fish catch. A number of factors, including an escalating world population coupled with blatant overfishing, have resulted in dramatic decreases in coastal zone food-fish stocks on a global basis. Some stocks (such as North American Pacific salmon and Atlantic cod) are virtually depleted. A number of factors, including contaminant-related endocrine disruption, have decimated inland water food-fish stocks.

Understandably, therefore, management of human activities in inland and coastal waters has become a recognized priority on a global scale (e.g., International Geosphere–Biosphere Program, national and regional programs of monitoring and assessing quality of water for consumption, conservation, industrial aquaculture, and recreation). Incorporating remotely generated inland and coastal water quality products into this recognized global priority clearly exemplifies value-added remote sensing.

Over the past three decades, scientists from government, academic, and private sector research institutes have developed bio-optical and bio–geo-optical models and algorithms that have enabled the monitoring of inland and coastal water quality from space. It is now possible to use historical and current satellite information, along with data from new environmental satellites equipped with more sophisticated spatial and spectral resolution, to detect and monitor environmental trends in lakes and rivers on local to global scales. Remotely sensed water quality mosaics may now be used to assess impacts of environmental stresses on the Earth's standing water resources.

Despite the prolonged and prejudicial beliefs by much of the ground-based monitoring community that replacement is the intent of proponents of environmental remote sensing, monitoring by remote sensing can never replace ground-based water quality monitoring. However, it has many potential applications to help protect the environment and benefit the economy. Early indications of ecosystem damage and prompt corrective action can increase sustainability of forestry, agriculture, and resource development. Resource managers can evaluate the success of restoration programs aimed at improving water quality — a definite advantage at a time when the world is becoming increasingly concerned about the quality of its water.

The great discoveries of science often consist, as we saw, in the uncovering of a truth buried under the rubble of traditional prejudice, in getting out of the cul-de-sacs into which formal reasoning divorced from reality leads; in liberating the mind trapped between the iron teeth of dogma.

From *The Sleepwalkers*, **Arthur Koestler, 1959**

chapter five

Inland and coastal (case 2) water quality products

Trained in college to believe that to look beyond the immediate moment — to look for causes or to foresee consequences — is impossible, modern men have developed context-dropping as their normal method of cognition. Observing a bad, small-town shopkeeper, the kind who is doomed to fail, they believe — as he does — that lack of customers is his only problem; and that the question of the goods he sells, or where those goods come from, has nothing to do with it. The goods, they believe, are here and will always be here. Therefore, they conclude, the consumer — not the producer — is the motor of an economy.

From *Philosophy: Who Needs It?* Ayn Rand, 1982

Throughout this book, we have considered the plight of remote sensing of inland and coastal waters, the "context" of which is that its products face a distinct lack of customers outside the scientific community. Remote sensing is a data-gathering tool observing effect rather than cause. Influential environmental end-users of these data are primarily concerned with cause. Although cause and effect are inescapably related, these relationships are not always immediately and/or intuitively obvious. Scientists are effective at deducing and explaining cause–effect relationships, but they are notoriously ineffective at marketing their work. The private sector community will, quite sensibly, follow the directives of funding agencies.

Aquatic science and remote sensing technology have generated inland and coastal water quality products that benefit environmental monitoring and should, therefore, have a consumer interest. It thus behooves the producer to become the motor of the

economy by making funding agencies see that the contents of
these products are worthy of support. In Chapter Four, we out-
lined some roles that remote sensing in general and remote sensing of inland and coastal waters in particular might play in
environmental stewardship. In this chapter, we present some (although as yet limited) examples that will hopefully illustrate these
roles. Applications of remote sensing of inland and coastal waters
are directed towards the protection and improvement of human
and ecosystem life, health, and safety. Thus, for the scientific
remote sensing community to obtain an end-user market for its
products via government and/or space agency funding bodies,
it must effectively make public good become a private sector
product as well as a marketable commodity.

5.1   Providers and users: a marketplace reality check

Much focus has been directed herein and elsewhere towards the marketplace
(or lack thereof) for inland and coastal ocean water quality products. A large
number of global environmental programs have been established (and more
are being considered) with driving forces that include (1) improving our
understanding of local, regional, and global processes and their controlling
roles in environmental and climatic change; and (2) predicting the potential
impacts of natural processes and anthropogenic stressors on society. A list
of such global programs would include the United Nations' Global Ocean
Observing Systems (UN-GOOS) and Global Terrestrial Observing System
(UN-GTOS); the World Meteorological Organization's World Climate
Research Program (WMO-WCRP) and Global Climate Observing System
(WMO-GCOS); and the Land–Ocean Interactions in the Coastal Zone as a
component of the International Geosphere–Biosphere Program
(LOICZ-IGPB), among others. Recently, an international Group on Earth
Observations (GEO) has been charged with establishing an international
team approach to building an Earth Observation System (EOS) that would
incorporate ground-based and space-based monitoring systems on national
and international bases.

The nobility of the goals of the preceding organizations is laudable,
although specific differences among the goals may be difficult to discern.
Each of the global programs tacitly assumes that current and future
space-based monitoring activities will continue and will play prominent
roles. However, whether consciously or unconsciously, all the global programs are unfortunately predicated upon two common provocative initial
postulates that subliminally subscribe to the Ayn Rand economic motor force
philosophy (as expressed in the opening and closing quotations of this chapter), namely:

- Current established or pursued nonscientific end-users of ground- or space-based EO data products do not generally regard providers of these products as bearers of marketable goods.
- Current providers of ground- or space-based EO data products do not generally regard the current established or pursued nonscientific end-users of these products as serious customers for their marketable goods.

These two subliminal postulates are consistent with the retrospection and introspection presented throughout this book. They do, however, provide a reality check for both the providers and users of EO data products and should emphasize to them the necessity of a marketplace existing for their common good. Neither the pursued users nor the providers of EO data products regard the others as an economic motor force of the EO marketplace and the fate of the producers of EO data products is completely dictated by the profitability of the marketplace; therefore, it is incumbent on the providers to become the motor force of the marketplace and its economy.

As part of this reality check, some mention should be made of a not infrequent pitfall associated with the search for and implementation of international markets. It has become common practice for governments of developed nations to assist governments of developing nations in a variety of humane areas such as health, food production, shelter, mechanization, economic and political self-sufficiency, clean air, potable water, farming practices, and many other life-improving needs. Quite apart from altruism, developed nations regard such assistance as a means of establishing international markets for their manufactured goods. Satellite-derived environmental products and the methodologies responsible for them fall into the category of manufactured goods, with environmental assessment and clean-up as valid justifications for seeking these international markets (as an example, products of the Canadian satellite Radarsat-1 are used for and in a number of developing nations).

Without intentional denigration of such remote sensing assistance and market development, however, it is necessary to underscore certain realities:

- As we have discussed at various points in this book, remote sensing is an expensive undertaking with the costs of airborne flights, historic and current satellite data, interpretation facilities, and obligatory ancillary fieldwork beyond the financial envelopes of many, if not most, research and educational institutes in developed nations.
- Developing, nondespotic nations do not have the financial resources or the expertise or facilities of developed nations.
- As we have discussed throughout this book, the scientific and technical developers of these products are their principal end-users.

- As a consequence of the preceding three points, the developed nations invariably provide all the scientific and technical expertise, facilities, and financial resources to establish this market.
- A cycle is then very often set in place that roughly runs according to a certain scenario.

In the scenario mentioned in the last item, the government of a developed nation establishes, at public expense, a satellite program (design, development, launch, tracking, data collection, data dissemination). The government then awards contracts, at public expense, to government, academic, and private sector scientists to generate remote sensing products that often, as in the case of water quality products, lack a dedicated national market. The government then awards contracts, at public expense, to the private sector to manufacture these products for international sale. (It should be added that a similar scenario applies to hardware and software products such as sensing devices, work stations, and data capture engines required for on-site downloading of real-time data. Again, the cost is passively borne by the tax-paying public of the developed nation.) The government then provides environmental data products, facilities, and work stations to the developing nations through direct aid, interest-free loans that then become principal-free loans, and incentive payments to heads of state, again at public expense. To provide bizarre closure to the cycle, the government sends, again at tax-payers' expense, remote sensing experts to work with and train uninformed, and generally uninterested, local citizens to analyze and use these environmental products.

Thus, this approach to obtaining a vibrant and effective international market rapidly becomes an incestuous merry-go-round whereupon a developed nation uses its tax dollars to generate environmental products and then pays developing nations more of its tax dollars to buy these products (in which national and international users often lack interest). Such a practice gives the appearance of establishing an international market, thereby allowing the merry-go-round to continue its unabated rotation. An added horror to this operation is that once interest can no longer be feigned within one developing nation and the merry-go-round is moved to another, all the sophisticated high-tech remote sensing data analyses facilities and data capture hardware and software remain unused and unattended at the first foreign site, only to be duplicated then at the new site. (This is much to the chagrin and envy of the scientific community of the developed nation that finds attainment of expensive remote sensing facilities beyond its restricted budgets.)

Such an enticing merry-go-round philosophy can be quite seductive. However, paying twice, thrice, or many more times for products or services of questionable market interest is an urgent issue that must be seriously addressed. One obvious way is to illustrate first the value of the products through the establishment of a national market of influential end-users. To do so requires that the sponsoring tax-paying public be convincingly brought on side. Therefore, we reiterate that the scientific remote sensing community must effectively make public good become a private sector product and a marketable commodity.

...what fools these mortals be!

From *A Midsummer Night's Dream*, Act 3, Scene 2, William Shakespeare, 1595

5.2 Applications of inland and coastal water color products

Lake and coastal ocean environments are influenced by a variety of physical drivers such as winds, runoff, tides, and bathymetry; a variety of chemical drivers such as direct and indirect injection of elements and compounds of natural and anthropogenic origin; and a variety of biological drivers such as populations of indigenous terrestrial, aquatic, and atmospheric species of domesticated and undomesticated life forms. Such dynamic ecosystems are governed by events and processes that operate on time and space scales that are small relative to those events and processes (apart from catastrophic step-function) governing the open ocean.

Due to its ability to view readily accessible and inaccessible (by surface monitoring stations) regions of the Earth's surface, remote sensing can and should be considered as a stand-alone monitoring tool when so required or as a component of an integrated monitoring network when *in situ* monitoring occurs in tandem. Thus, for some geographic locations, remotely sensed water quality products must be considered in the absence of ground-acquired environmental information. However, for the inland and coastal waters that support the bulk of the Earth's human population, remotely sensed water quality products are generally not stand-alone entities, but rather complementary to environmental products resulting from directly acquired data. To optimize the benefits of such complementarity, efforts must be accelerated to develop and employ integrated approaches to environmental monitoring that assimilate remotely sensed water quality (and *in situ* optical) products and ground-acquired environmental products into diagnostic and prognostic assessments of environmental behavior.

In addition to providing information on the physical and geophysical aspects of inland and coastal ecosystems (e.g., hydrology, water quantity, dynamics, near-surface temperature, shoreline features), remote sensing of water (specifically color) provides information on the nature of the water (e.g., water clarity, biomass, coextant concentrations of phytoplankton, dissolved organic matter, and suspended sediment). Examples of remotely sensed water quality products providing such information are presented and discussed in this chapter. Unfortunately, however, we can draw such examples only from a limited catalogue. Certainly the Internet has been a boon to accessing at least recent satellite images collected over the globe, and some of these images are over inland and coastal water bodies. Most images are direct or enhanced "pictures" of the monitored region (i.e., essentially level 1 data products), although level 2 and some level 3 products are increasingly available (e.g., chlorophyll, suspended sediments, attenuation coefficients).

A variety of selected operational models is routinely used to generate these higher level products. As discussed in Chapter Three, removal of the atmospheric intervention can be quite problematic. Also, as discussed earlier, the generation of water quality products falls into the domains of research, model development and validation, and reassessment of performance and properties of existing and planned satellite sensors. Thus (admittedly by necessity as well as possibly by choice), we will select examples from established, ongoing, or recent research projects. Readers, of course, are encouraged to surf the Internet for illustrations of maps of chlorophyll and sediment concentrations from Web sites of space agencies, limnological and oceanographic institutes and organizations, academic institutes, and private sector companies specializing in remote sensing and remote sensing products.

The majority of the needs of environmental decision-makers and policy-makers that were discussed in Chapter Four depend on accessing time-series of remotely sensed data (a recurrent theme of this book). Over three decades of impressive archival satellite data already exist; this archive continues to burgeon with space-acquired data of increasing spectral sensitivity. Sequential satellites soon to be launched by major space agencies, combined with programs aimed at intercalibration of their acquired data, will further extend these archives and provide long-term and consistent data sets that are not possible from any other monitoring system. Inland and coastal water quality products emergent from these data sets are capable of dramatically enhancing sound environmental stewardship.

5.3   Mapping water clarity

"Clarity," like its generally considered aquatic antonym, "turbidity," is a somewhat loosely defined term describing aquatic composition. Both terms refer to the totality of suspended organic and inorganic matter coextant in the water column, but not in solution. Both are indicative of the transparency of natural water to visible radiation. Thus optical transmission becomes a measure of water clarity. Waters of high clarity (i.e., high optical transmission) may be confidently considered to be relatively free of suspended particulates. Waters of high turbidity (i.e., low optical transmission) possess substantial concentrations of suspended particulates. The demarcation between clarity and turbidity is subjective and artificial.

Water clarity is not a measure of water quality because waters with predominantly organic suspended particulates possess a quality considerably different from waters whose suspended particulates are predominantly inorganic. Furthermore, suspended particulate (contaminated or uncontaminated) concentrations could produce a significant impact on light attenuation. However, interstitial chemicals (pernicious or benign) in aquatic solution, irrespective of concentrations high enough to warrant serious consideration, cannot attenuate light in a manner that would rival the scattering and absorption produced by chlorophyll, suspended minerals, and dissolved organic carbon. Ultimate water clarity (water molecules) represents aquatic barrenness. Lake Tahoe in Nevada and Lake Superior along the Canada–U.S. border are illustrations of barren and relatively barren inland waters, respectively.

Precise measurements of *in situ* optical transmission may be readily obtained by transmissometers. Historically, however, optical clarity has been reported by use of a Secchi disk, named after its acclaimed founder (Secchi, 1866). This disk (generally a round plastic or metal object of color ranging from white to off-white to patterns of alternating white and black segments) is distinctive enough to facilitate recognition by the human eye and is lowered into deep waters and visually tracked until it disappears from the view of an observer located above the air–water interface. The depth (in meters) at which this disappearance is deemed to occur is termed the *Secchi depth*. The greater the observed Secchi depth, the clearer the water.

It is easy to criticize the use of Secchi disks (nonuniformity of the disks, the large number of observers with nonstandardized eyesight, nonstandard viewing and solar angles, and wide variability of climatic conditions under which observations are taken) as a measure of optical clarity. However, despite these limitations, a variety of workers (Duntley and Preisendorfer, 1952; Beeton, 1957; Tyler, 1968; Gordon and Wouters, 1978; Bukata et al., 1988b, among others) has shown that the Secchi depth provides a quantitative estimate of a *hybrid* optical property, comprising, in part, an *inherent* optical property (the beam attenuation coefficient, c, which is the sum of the bulk absorption coefficient, a, and the bulk scattering coefficient, b) and in part by an *apparent* optical property (the irradiance attenuation coefficient, K_d). Thus, Secchi depth may be regressed as a function of $(c + K_d)$, that is, as a transparency that is a dual consequence of the composition of the water body and the spatial configuration of the incident and in-water light fields.

However, for such lake-dependent relationships to emerge, a very large number of observations are required. Statistical scatter must be overcome by sheer volume of data. Nevertheless, in the absence of more rigorous data sets, the cautious use of Secchi depths can benefit water managers and decision-makers. Time series indicating changes in water clarity can provide insight into aquatic ecosystem change, particularly when such Secchi depth data may be used in conjunction with *a priori* environmental information (such as changes in basin land use, invasion by exotic species, changes in agricultural and/or mining practices, etc.). The longevity of the Secchi depth database (for many natural water bodies, Secchi depths represent the sole recorded optical history) and the appropriately ascribed convenience and deceptive simplicity of its usage attest to this disk's popularity within the limnological science community.

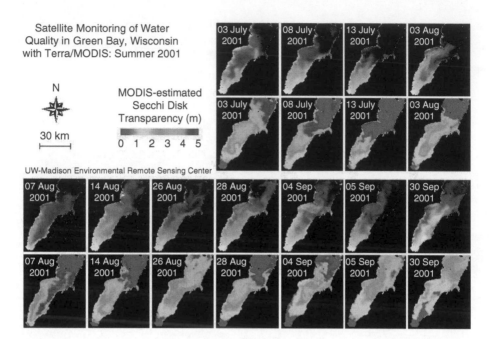

Figure 5.1 (See color insert following page 134.) Time series of Secchi disk transparency maps generated from Terra/MODIS images of Green Bay, Wisconsin, over the summer of 2001. (From Richardson, L.L. et al., *Backscatter*, winter issue, 26–31, 2002.)

Generation of Secchi depth "maps" from space data may be of considerable value to environmental managers concerned with changes in inland water clarity. Landsat images are currently used in such a manner to record region-wide changes in water clarity in Minnesota, Wisconsin, and Michigan (Chipman et al., 2004) as part of the Wisconsin Satellite Lake Observatory Initiative (SLOI) (Lillesand et al., 2001). Figure 5.1, taken from Figure 5 of Richardson et al. (2002), illustrates results from the joint Green Bay water quality project involving the UW-Madison Environmental Remote Sensing Center, the North Temperate Lakes Long-Term Ecological Research Site, the Wisconsin Department of Natural Resources, the Upper Midwest Regional Earth Science Applications Center, and the NASA Affiliated Research Center Program. This figure depicts a time series of false color Secchi disk transparency maps generated from Terra/MODIS images for Green Bay, Wisconsin, in the summer of 2001. Spatial and temporal variations in optical clarity are dramatically illustrated.

The Secchi depths of Figure 5.1 are estimated through a regression model utilizing observed spectral radiation from a combination of satellite spectral bands and an extensive coordinated program of ground-based data collection involving almost 1000 citizen volunteers. In this case, the science is simple and lacks the mathematical rigor demanded of sophisticated bio–geo-optical models. Simplicity notwithstanding, the science is sufficiently sound to present convincing evidence of inland water *clarity* in a manner that is readily understood by and relevant to environmental decision-makers and political policy-makers. Therefore, such Secchi disk transparency maps satisfy the criterion for the compromise between defendable science and defendable usage of that science. They are also compatible with the suggestion made in the opening preamble to Chapter Four to regard open mindedly the possible need, under certain specific circumstances, to revisit use of the simple regressions involving mathematically manipulated ratios of remotely recorded radiances of differing wavelengths that work so well for open ocean waters. Finally, the need for a legitimacy seal from the influential end-user community is well served by the integral involvement of the public whose inland water concerns are being addressed by remote sensing.

5.4 Mapping chlorophyll concentrations (and primary production) of mid-oceanic (case 1) waters

Photosynthesis is the biological combination of chemical compounds in the presence of light. It generally refers to the production of organic substances (primarily sugars) from the carbon dioxide and water residing within green plant cells, provided the cells are sufficiently illuminated to enable plant chlorophylls to transform the radiant energy into an organic compound. Natural and cultivated regional flora provide the principal photosynthetic organisms to terrestrial biomes. Regional benthic and planktonic algae (and higher forms of plant life) provide the principal photosynthetic organisms to aquatic biomes. *Primary production* is the total chemical energy contained within an ecosystem as a direct result of photosynthesis. *Primary productivity* is the sum of all photosynthetic rates within an ecosystem. Assessment of aquatic primary production, therefore, requires the daily integrated photosynthesis occurring within the water column.

Because phytoplankton are the principal photosynthesis agents of case 1 waters, mid-ocean primary production may be confidently considered as phytoplankton photosynthesis. Aquatic photosynthesis is determined from *in situ* measurements of oxygen production or carbon (^{14}C) uptake at a variety of depths. Aquatic primary production is generally recorded in units of grams of carbon per square meter per day (or per month or year, depending upon the specific time-series variability sought). In order to qualify as a unit of measurement for living matter, *phytoplankton biomass* must be related to a biological parameter such as chlorophyll, carbon, or nitrogen. Consistent with Strickland (1965), Platt and Irwin (1973), and others, we define phytoplankton biomass as the organic carbon content of the phytoplankton population.

Estimation of aquatic photosynthesis requires knowledge of three relationships (Smith, 1936; Talling, 1957; Fee, 1969, 1990; Vollenweider, 1965; Bannister, 1974; Platt, 1986; Sathyendranath and Platt, 1989a, among others):

- Solar photosynthetic available radiation (PAR, wavelength region 400 to 700 nm) as a function of time, i.e., E_{PAR} vs. t
- Photosynthesis, P, as a function of PAR, i.e., P vs. E_{PAR}
- PAR as function of aquatic depth, i.e., E_{PAR} vs. z

PAR represents the totality of solar radiation available for photosynthetic activity. However, the absorption spectra of algae are not invariant over the 400- to 700-nm wavelength interval. Thus, all of PAR does not result in photosynthesis. The portion of PAR pertinent to photosynthetic activity is termed the photosynthetic usable radiation (PUR) and is mathematically calculable as a weighted value of PAR. Relationships between PUR and PAR for the Laurentian Great Lakes are given in Jerome et al. (1983).

Although the first relationship may be readily obtained from direct measurements immediately above the air–water interface, the second and third relationships have formed the subject of volumes of scholarly works. Obtaining these relationships is not a simple task for operational water quality monitoring from space vehicles. We shall not pursue these complexities here. More expanded discussions of remote sensing and bioproduction may be found in Platt and Herman (1983); Sathyendranath and Platt (1989b); Bukata et al. (1995); and elsewhere. Suffice here to say that the Coastal Zone Color Scanner (CZCS) launched aboard Nimbus 7 in 1978, which remained operational until 1986, enabled assessments of oceanic distributions of chlorophyll concentrations and phytoplankton biomass (Putnam, 1987; Müller–Karger et al., 1989).

Remotely monitoring phytoplankton biomass (organic carbon content) requires a relationship between chlorophyll concentration and the organic carbon content of algae — a relationship that not only depends on algal species, but also varies considerably in space and time. Remotely estimating the organic carbon content of phytoplankton in a natural ocean water environment free of the obfuscating presence of detritus and zooplankton, however, is virtually impossible. (In inland and coastal waters the obfuscating presence of additional CPAs produces an even greater impact.)

Therefore, in order to estimate aquatic biomass from a remote determination of chlorophyll concentration, a derived or postulated carbon-to-chlorophyll ratio is required. Because such a ratio is a highly elusive commodity, a generally accepted practice has been to consider remote estimates of near-surface chlorophyll concentrations as acceptable surrogates for estimates of near-surface biomass. Regarding the near-surface chlorophyll concentration as a first-order estimate of phytoplankton biomass is reasonable and consistent with the compromise between science and usage of science for applications advocated throughout this book. Further valuable detailed discussions are to be found in Eppley and Peterson (1979); Harris (1980); Bannister and Laws (1980); Eppley et al. (1985); Campbell and O'Reilly (1988); Platt and Sathyendranath (1991); and Howard and Yoder (1997), among others.

The Bedford Institute of Oceanography (BIO) in Dartmouth, Nova Scotia, and Dalhousie University in Halifax, Nova Scotia, have made epic contributions to the science of monitoring ocean chlorophyll and primary production from space (see, for example, Platt et al., 1989, 1991; and Sathyendranath et al., 1995, among others). Taken from the Nova Scotia work of Longhurst et al. (their Figure 2; 1995), Figure 5.2 illustrates a global annual mean field of primary production in the *photic zone* of the oceans in units of $gCm^{-2}y^{-1}$. (The photic zone, also referred to as the *euphotic zone*, is the aquatic region in which maximum photosynthesis occurs, generally taken to be the upper aquatic layer bounded by the 100% and 1% subsurface irradiance levels. The vertical distance between these irradiance levels is referred to as the *photic depth* or *euphotic depth*.)

The global field of oceanic primary production of Figure 5.2 was determined from the monthly mean near-surface chlorophyll concentrations from 1979 to 1986 obtained by the Nimbus 7 CZCS through a meticulous application of the local primary production model and algorithm of Sathyendranath et al. (1995). As discussed earlier, the model required information of the spatial variability of the near-surface chlorophyll concentration and governing parameters of photosynthesis–light relationships, as well as atmospheric aerosols and Sun angle. Partitioning global ocean waters into 57 biogeochemical oceanographic subset regions (Platt and Sathyendranath, 1988) produced the primary production map of Figure 5.2.

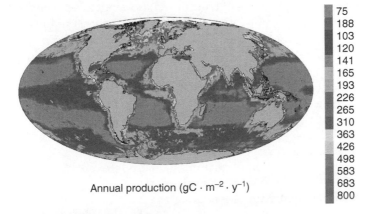

Annual production (gC · m⁻² · y⁻¹)

75	
188	
103	
120	
141	
165	
193	
226	
265	
310	
363	
426	
498	
583	
683	
800	

Figure 5.2 (See color insert following page 134.) Annual mean field of total primary production of phytoplankton in the euphotic zone of the oceans in gC m⁻² y⁻¹. (From Longhurst, A, et al, *J. Plankton Res.*, 17, 1245–1271, 1995.)

Exclusive of the coastal regions, the global net primary production from oceanic phytoplankton was estimated as 45 to 50 $GtCy^{-1}$ (compared with published estimates of 45 to 68 $GtCy^{-1}$ for global land vegetation and ~1.9 $GtCy^{-1}$ for coastal vegetation). Clearly, primary production maps such as Figure 5.2, along with dissolved organic matter (carbon) concentrations that can emerge from inverse modeling and multivariate analyses of remotely monitored coastal water color, are valuable products for understanding local and global carbon cycle dynamics. For further information, see the work of Berger et al. (1987).

5.5 Mapping chlorophyll concentrations (and primary production) of optically complex inland and coastal (case 2) waters

Successful mapping of aquatic chlorophyll concentrations generates a water quality product that is

- Inherent to the health and welfare of coexisting human and wildlife populations dependent upon sustainable shared water usage in a freshwater basin or a coastal ocean zone
- Comparable with terrestrial space products generated for estimating vigor and productivity of forests, farmlands, and grasslands
- Essential for incorporation into monitoring protocols of existing ground-based water quality monitoring networks
- Almost totally neglected by the end-usership for which this product is intended

Extracting concentrations of chlorophyll (in tandem with concentrations of its coextant CPAs) from water color has been the goal of bio–geo-optical modeling of case 2 waters since the 1970s and has been the major focus of this monograph. As we have discussed, numerous methodologies may extract and map these chlorophyll concentrations from space-acquired spectra of aquatic color. Here we have endorsed multivariate optimization and forward-inverse models involving *in situ* water sampling in concert with *in situ* optical measurements. Nonetheless, other endorsements could be equally valid. However, numerous *empirical* (reliance solely upon observation and practical experience rather than well-established scientific principles) chlorophyll-extraction algorithms (generally expressed as power law, multiple regression, hyperbolic, cubic, series, and other mathematically defined relationships) have been developed by numerous workers from numerous scientific institutes and are routinely used by space agencies to generate level 2 and level 3 chlorophyll concentration products.

Empiricism, however, is not necessarily "a bad thing" because such algorithms, if deemed appropriate, might fit quite snugly within our advocated compromise between defendable science and defendable usage of science (despite the acknowledged disparities of empiricism and well-established science). However, such empirical (and/or quasianalytic) chlorophyll-extraction algorithms are constrained to be functions of the discrete spectral wavelengths that are remotely monitored. Thus, the empirical algorithms become unique to each satellite sensor.

Furthermore, a wide choice of mathematical formulism presents itself to empiricism. Therefore, a variety of empirical algorithms becomes available for consideration for generation of level 2 chlorophyll products for the same satellite sensor data. It would be informative at this point to illustrate an example of how this repetitive profusion of emergent (and/or historical) empirical chlorophyll-extraction algorithms is addressed. Obviously, clever intercomparison and definitive acceptance–rejection criteria are required. Such scrutiny, however, invariably underscores shortcomings in specific empirical algorithms and also exposes problems inherent to the usage of *any* chlorophyll-extraction algorithm or methodology — particularly for the optically complex case 2 waters. The uncompromising nature of natural waters once again assumes its role of a limiting factor. (These data intercomparison difficulties are the targets of "binning" schemes referred to in the sidebar in Section 6.1.)

O'Reilly et al. (1998) presented an expert critique of the ocean color chlorophyll-extraction algorithms considered for use with the SeaWiFS sensor. A large data set containing coincident remote sensing reflectance (R_{RS}) values and *in situ* chlorophyll concentrations was compiled from 919 stations. Over 98% of the stations were in nonpolar case 1 waters; the remainder were in case 2 coastal waters. A variety of statistical and graphical criteria was then used to evaluate the appropriateness of 2 semianalytic and 15 empirical chlorophyll-extraction algorithms in replicating the compiled (chlorophyll, R_{RS}) data set. Differences in algorithm performance were in evidence and a hierarchy of appropriateness was established.

Despite the obvious value of such intercomparisons, however, an endemic reality of local application of these algorithms (regardless of hierarchy) also emerges. Collectively, the 919 stations displayed over three orders of magnitude of chlorophyll concentration values. The vast majority of values were in the range ~0.06 to 0.7 µg L^{-1} (with a geometric peak at ~0.2, a value close to that of the global ocean mean). Considerably smaller peaks were centered at ~0.03, ~1.0, and ~5.0 µg L^{-1}, respectively. A very limited number of noncontinuous values extended the chlorophyll concentration range to ~40 µg L^{-1}. Convincing scatter plots of R_{RS} vs. chlorophyll concentration (r^2 values from 0.917 to 0.876) were obtained for *all* the algorithms tested when data pairs from the 919 stations were utilized.

Clearly, however, these correlations were weighted by oligotrophic case 1 oceanic waters where chlorophyll is the sole aquatic CPA and the numerical range of chlorophyll concentrations is generally well within a single order of magnitude. Furthermore, the correlations also incorporated two additional orders of magnitude extending the data points to include mesotrophic and eutrophic waters where suspended inorganic matter and dissolved organic matter are additional CPAs that often overwhelm chlorophyll as a dominant CPA, and the numerical range of chlorophyll concentrations may be well within or beyond a single order of magnitude. Obviously, the O'Reilly et al. work is invaluable in illustrating why empirical algorithms have displayed such success in mapping oceanic chlorophyll concentrations and also for establishing a hierarchy among them. It must be remembered, however, that such convincing scatter plots emerged from a consideration of data from 919 sources. (Actually, an initial data set from 973 sources was considered but "outlier" data from 54 suspect stations were rejected.) Such emergent relationships are reminiscent of the Secchi depth relationships with photon attenuation briefly discussed in Section 5.3 in which the statistical scatter required being overcome by sheer volume of data before the relationships could be seen.

Most local inland (and some coastal ocean) waters do not have the luxuries of large ranges of chlorophyll concentrations, known or readily determinable concentrations of nonchlorophyllous CPAs, or (R_{RS}, chlorophyll, CPA) data sets large enough to overcome statistical scatter. Apart from spring plankton blooms or off-season deleterious plankton blooms, most inland water chlorophyll concentrations lie within one or two orders of magnitude and often within only a factor of two or three. Additional optical complexities, limited access to directly relatable large data sets, and accuracies of ~ ±30% generally attributable to remotely estimating chlorophyll concentrations of inland waters combine to make such an intercomparison of NASA level 2 algorithm usage a difficult, if indeed not possible, venture for inland waters (this prior to consideration of the major obstacle of atmospheric obfuscation of the aquatic target).

Difficulties in evaluating level 2 chlorophyll algorithms for inland waters are being scientifically addressed within the NOAA CoastWatch Program (G.A. Leshkevich and B.M. Lesht, private communication); the Canadian Space Agency's evolving and updating level 2 Algorithms Theoretical Basis Document (ATBD) for the MERIS sensor; the Naval Research Laboratory's Ocean Science Branch mandate; and elsewhere. These and other considerations emphasize this monograph's endorsement of the use of multivariate optimization and analytic bio–geo-optical models for space EO of local inland water quality. Although empiricism appears to be a valid tool for case 1 waters, it needs to be used carefully, and perhaps sparingly, for case 2 waters.

All water quality models require realistic values of the inherent optical properties of the indigenous CPAs as well as realistic values of the *atmospheric attenuation coefficient,* the *aerosol optical depth,* and the *aerosol phase function* (see the glossary in Appendix B for definitions) pertinent to the atmospheric column separating the remote sensor from the targeted water body. All water quality models are local in application and all water quality models (some more sensor dependent than others) might produce value-added water quality products satisfying the criteria of "defendable science vs. defendable usage of that science."

With the preceding interpretation caveats, the restricted ranges of inland water CPA concentrations, and atmospheric obfuscation issues firmly in mind, however, illustrations of chlorophyll patterns synoptically observed in an interconnected lake system may be amenable to science-application compromise. Figure 5.3 (provided courtesy of G.A. Borstad Associates, Ltd) illustrates a pseudocolor April 26, 2003, SeaWiFS chlorophyll concentration image of the Laurentian Great Lakes. Rotating counterclockwise from the 12:00 position are the three upper lakes Superior, Michigan, and Huron with Georgian Bay, and from west to east are the two lower lakes Erie and Ontario. The image (1-km pixels) was constructed from two consecutive scenes utilizing NASA operational empirical levels 2 and 3 water quality algorithms. Image enhancement was generated from a logarithmic color bar that was equalized by number population rather than linearly allowing "swirls" to show in the data. White in the image is "flagged" (i.e., removed from the data stream as being cloud or ice), as is black (i.e., missing data or something other than cloud or ice).

Interestingly, as well as frustratingly, Figure 5.3 displays acceptable and unacceptable features as chlorophyll concentration patterns. Although the lake patterns in lakes Superior, Huron/Georgian Bay, and Michigan could be consistent with expected early spring chlorophyll distributions, the "chlorophyll" distributions in lakes Erie and Ontario are almost certainly due to

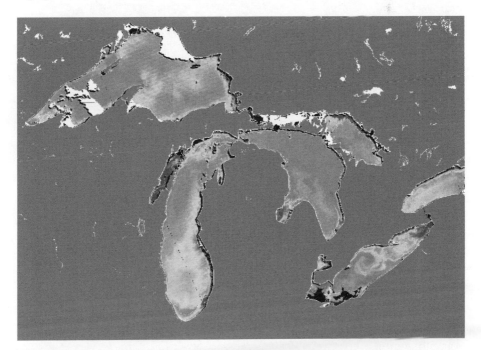

Figure 5.3 Pseudocolor April 26, 2003, SeaWiFS chlorophyll concentration image of the Laurentian Great Lakes. (Provided courtesy of G.A. Borstad Associates, Ltd.)

inorganic matter. Clearly, a time-series of images (weekly, monthly, seasonal, yearly) like Figure 5.3 can be very valuable in delineating inland water quality change. It can, however, be very misleading or totally wrong. There-fore, extreme caution must be exercised in the use of appropriate time-series data; *a priori* knowledge of aquatic behavior is most often the only acceptable criterion upon which to evaluate acceptability of space-acquired data (again demonstrating the need to merge space and ground-based data sets).

Figure 5.4a (taken from Figure 3 of Gower, 2004) illustrates NASA time-series SeaWiFS monthly averages of Great Lakes chlorophyll concen-trations (μg L^{-1}) determined in a similar manner as for Figure 5.3 but with a slightly different color scale. Shown are mean monthly chlorophyll con-centration images extracted from the standard mapped images for 1997 to 2001, inclusively. Figure 5.4b illustrates the normalized water-leaving radi-ance (W m^{-2} sr^{-1} · μm^{-1}) at 555 nm taken from the same time-series data set. As in Figure 5.3, flagged data in both sequences are displayed in black. Despite the reduced scale of Figure 5.4, distinct temporal changes in the lake-wide patterns of each lake are easily seen. Again, as for Figure 5.3, extreme caution coupled with *a priori* knowledge of local lake behavior is required to assign acceptable and unacceptable meaning to the surficial patterns in each time-series.

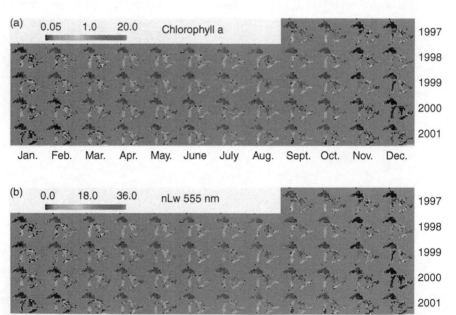

Figure 5.4 (See color insert following page 134.) Time-series SeaWiFS data water quality features collected over the Laurentian Great Lakes: a) average chlorophyll concentration; b) average water-leaving radiance at 555 nm. (From Gower, J.F.R., *Can. J. Remote Sens.*, 26–35, 2004.)

The water quality in the Great Lakes manifests as a complex interdependency of local geologic, anthropogenic, and climatic parameters. For example, most, if not all, of the "chlorophyll" patterns from November to February could be discarded due to extended snow and ice conditions in the Great Lakes. Subsequent to ice melt and spring thaws, the chlorophyll patterns in the upper lakes begin to develop in a meaningful manner. However, the high run-off and highly concentrated littoral drift within the lower lakes results in concentrations of organic matter becoming overwhelmed by concentrations of inorganic matter.

The intensity time-series maps of the red wavelength 555-nm water-leaving radiance essentially result from particulate CPA scattering centers within the upper attenuation length of the surficial water. Organic (chlorophyll) and inorganic (suspended minerals) CPAs scatter red wavelengths. Thus, red wavelength patterns can confound organic and inorganic concentrations into an inseparable turbidity map. As discussed in Chapter Two and Chapter Three, separation of chlorophyll from suspended sediments requires full-color optical models and multivariate optimization methodologies. For this reason we have always advocated use of full-color water spectra of high spectral sensitivity to interpret such bright-water (i.e., high scatter at 555 nm) events as sediment transport, thermal upwelling, calcium carbonate precipitates ("whitings"), plankton blooms, and input plume dynamics. However, circumstances alter cases and sometimes the water-leaving radiance in the red region of the visible spectrum can serve as a stand-alone data source and provide aquatic information that might satisfy the obligatory compromise between defendable science and defendable usage of that science. More on the use of red wavelengths appears in Section 6.7.

Careful analyses of time-series data such as Figure 5.3 can provide information on a variety of episodic events, seasonal cyclic aquatic behavior, and inter-annual variability of inland water quality for any region of the globe. The Bedford Institute of Oceanography (BIO) routinely produces time-series of SeaWiFS data (chlorophyll and temperature) for coastal waters of eastern Canada. The Great Lakes Environmental Research Laboratory (GLERL) routinely produces time-series of SeaWiFS data for the totality of Great Lakes waters. The Institute of Ocean Sciences (IOS) routinely produces time-series of SeaWiFS data for the coastal waters of Western Canada. Linking such aquatic time-series data with those from nearby or adjacent terrestrial regions greatly assists the development of regional environmental impact assessment models.

Large continental coastal regions such as gulfs, straits, and bays contain admixtures of ocean and inland waters and often display optical complexities somewhat middling to those of archetypical case 1 and case 2 waters. Figure 5.5, courtesy of Robert A. Arnone, illustrates the chlorophyll concentration distribution in the Gulf of Mexico as recorded by SeaWiFS on July 14, 2004. The NASA OC4 chlorophyll extraction algorithm was applied to the water-leaving radiance once atmospheric intervention was removed by the Naval Research Laboratory (NRL) atmospheric correction model.

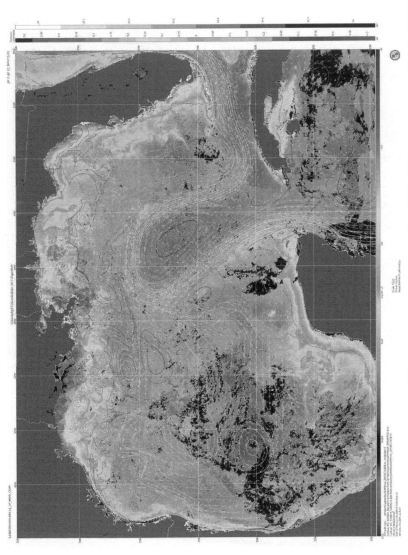

Figure 5.5 (See color insert following page 134.) Chlorophyll concentration in the Gulf of Mexico as recorded by SeaWiFS on July 14, 2004, using the NASA OC4 chlorophyll extraction algorithm. Also shown are contoured sea surface heights and salinity. (Provided courtesy of Robert A. Arnone, Naval Research Laboratory, Stennis Space Center.)

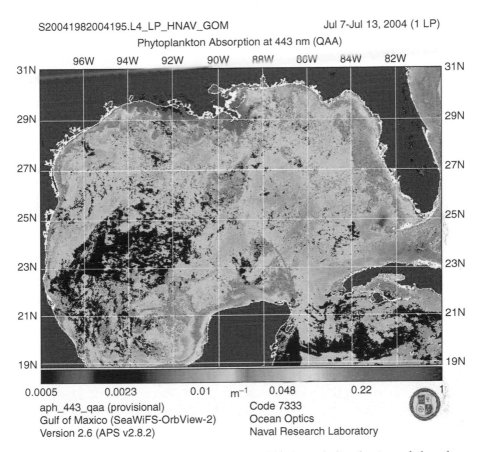

S20041982004195.L4_LP_HNAV_GOM   Jul 7-Jul 13, 2004 (1 LP)
Phytoplankton Absorption at 443 nm (QAA)

0.0005   0.0023   0.01   m^{-1} 0.048   0.22

aph_443_qaa (provisional)   Code 7333
Gulf of Maxico (SeaWiFS-OrbView-2)   Ocean Optics
Version 2.6 (APS v2.8.2)   Naval Research Laboratory

Figure 5.6 (See color insert following page 134.) Spatial distribution of the phytoplanktonic absorption coefficient at 443 nm, $a_{chl}(443)$, over the Gulf of Mexico averaged over the SeaWiFS data of July 7 to 13, 2004, using an empirical quantum adiabatic algorithm. (Provided courtesy of Robert A. Arnone, Naval Research Laboratory, Stennis Space Center.)

Overlaid on the chlorophyll map is the surface current with the contoured sea surface height (SSH) as determined from the Naval Coastal Ocean Model (NCOM). The black contours are NCOM salinity values.

On the NRL/Stennis Space Center Web site are numerous excellent examples of SeaWiFS/OC4 chlorophyll distributions in pelagic ocean and coastal ocean waters as well as distributions of ocean inherent optical properties (IOPs) as determined from its in-house empirical quantum adiabatic algorithm (QAA) established by Zhong Ping Lee and his collaborators (Lee et al., 1996a, b, 1998). Figure 5.6 illustrates the thus determined spatial distribution of the phytoplanktonic absorption coefficient at 443 nm, $a_{chl}(443)$, over the Gulf of Mexico averaged over the SeaWiFS data of July 7 to 13, 2004. Similarly, spatial distributions for the total absorption coefficient, $a(443)$, the dissolved organic matter absorption coefficient, $a_{DOM}(412)$, and the total backscattering coefficient, $b_b(555)$, are also readily available from the QAA.

Near-surface chlorophyll concentration distributions obtained from space measurements of inland water color are not as immediately free from skepticism as those obtained from space measurements of mid-ocean and some coastal ocean color. Nevertheless, acceptable chlorophyll concentration distributions of inland and coastal waters are being carefully obtained and analyzed by an increasing number of dedicated workers. The sad reality, however, is that converting remote estimates of near-surface chlorophyll concentration distributions into acceptable estimates of inland and coastal primary production must still be considered to be in relative infancy. Although applauding the insightful science and methodology that resulted in the ability to generate global ocean primary production "maps" of Section 5.4, we must once again recognize added difficulties arising from the optical complexities of land-dominated water regimes.

It has long been known (Gordon and McCluney, 1975) that, for any wavelength, λ, 90% of the optical signal recorded by a sensor viewing the ocean originates within the so-called *penetration depth* (defined as the inverse of the downwelling irradiance attenuation coefficient, $K_d(\lambda)$), which equates to a physical depth, z, at which the in-water downwelling radiation attenuates to ~37% of its value just beneath the air–water interface (i.e., within depth $z_{0.37}$). A considerable amount of ocean chlorophyll resides beneath the penetration depth. Thus, to convert near-surface chlorophyll concentration to a total concentration within the aquatic column being viewed, knowledge of the nonuniform vertical depth profile of chlorophyll concentration is required.

An extensive feature of the depth profile of oceanic chlorophyll is the *deep chlorophyll maximum (DCM)* as discussed by Cullen (1982) and others. The DCM is generally located beneath the penetration depth but contained within the photic zone, i.e., somewhere between $z_{0.37}$ and $z_{0.01}$. For most ocean waters, $z_{0.01}$ translates to depths of ~46 m. By comparison, the case 2 waters of the lower Great Lakes display photic depths of ~4 to 30 m, and the essentially case 1 pelagic waters of Lake Superior consistently display photic depths of ~40 m. For most ocean waters, the chlorophyll depth profile may be mathematically expressed as a Gaussian distribution centered at the DCM and superimposed upon a constant background. As the depth of the DCM increases, the ability of a satellite sensor to observe it decreases.

Thus, many remote estimates of total chlorophyll concentrations (and from there remote estimates of primary production) depend upon models involving vertical "bobbing" of such Gaussian-based distributions. The modeling works of Lewis et al. (1983); Platt et al. (1988); Sathyendranath and Platt (1989b); Morel and Berthon (1989); and others, while capitalizing upon the mathematical formulism of the preceding statement, involve sagacity that far transcends the simplicity that easily, but fraudulently, might be read into such a statement.

Unfortunately, the assumption of a mathematical Gaussian distribution defining a limnological chlorophyll profile is not always valid. A variety of vertical profiles ranging from completely uniform to completely random with a wide range of intermediate generic distributions is possible, depending upon the trophic status of the water body (nutrient dynamics), climatic conditions, and growth cycles of indigenous aquatic vegetation. Furthermore, as discussed in Section 5.4, to estimate phytoplankton biomass (organic carbon content) remotely, the elusive carbon-to-chlorophyll ratios of indigenous algae are required. This elusiveness has resulted in the use of remote estimates of near-surface chlorophyll concentrations as a generally accepted surrogate for remote estimates of near-surface phytoplankton biomass. Such an approximation has been deemed appropriate for mid-ocean waters where chlorophyll may be considered the sole CPA of ocean color. Such an approximation may indeed also be appropriate for inland and coastal waters. However, the added organic and inorganic CPAs in these waters, in addition to making the extraction of near-surface concentrations of chlorophyll a much more complicated task, might also decrease the ability of chlorophyll concentration to serve as a surrogate for phytoplankton biomass. (Among other reasons, this is due to the presence of terrestrially derived dissolved organic carbon as an independent and competitive CPA.)

The complexities of converting inland and coastal chlorophyll concentrations into estimates of inland and coastal primary production have not escaped the scrutiny of ground- or space-focused water quality modelers and experimental scientists. The *in situ* scientists and modelers have a distinct logistical advantage over the remote sensing scientists and modelers; however, the work-in-progress status of generating inland and coastal primary production products is proceeding well at universities and research institutes, as well as within space agency, government agency, military, and private sector laboratory and computer facilities on a global basis.

As testimony to the scope of complexity involved in generating a level 2 or level 3 primary production map product for case 2 water bodies, the necessary steps and functions involved may be briefly listed as:

1. Obtain the optical cross section spectra (inherent spectro-optical properties) pertinent to the chlorophyll, dissolved organic carbon, suspended sediments, and possibly the detrital matter indigenous to the inland or coastal water body targeted for water quality monitoring. This entails an on-site coordinated field program of direct measurements of the subsurface visible radiation field and collection of water samples for laboratory analyses. If such field deployment is not possible, surrogate optical cross spectra from comparable waters might be suitable.
2. Obtain the upwelling radiance spectrum (water color) immediately above the air–water interface from
   - Direct measurement, which is most preferable

- Applying a realistic atmospheric correction algorithm to remove atmospheric "color" from the raw level 1 spectroradiometric satellite data, which is less preferable and fraught with the difficulties elucidated in Section 3.7 and elsewhere in this book
- Using the water-leaving radiance values as recorded at the spacecraft subsequent to the algorithm and input algorithm values as decided by the space agency generating and disseminating the space data, which is by far the most frequently used radiance data

3. Extract the coexisting concentrations of chlorophyll, dissolved organic carbon, suspended sediments, and possibly detritus from the water-leaving spectroradiance using the pertinent optical cross section spectra and a multivariate methodology of choice.

4. Estimate the phytoplankton photosynthesis. This entails obtaining E_{PAR} as a function of time. Ideally this would be directly measured as part of the field logistics. However, E_{PAR} may be obtained from level 1 satellite data as part of the preceding second point if the satellite supplies downwelling top-of-atmosphere (TOA) solar and/or sky radiance from which the atmospheric correction algorithm could yield E_{PAR} incident at the air–water interface. Modeling this incident E_{PAR} through the air–water interface will yield the maximum PAR available for photosynthesis (recall PAR vs. PUR). The pertinent primary production transfer coefficients for the water body being considered would then need to be determined — an activity that ideally involves laboratory incubations followed by ^{14}C uptake and photosynthesis measurements (see Vollenweider, 1961; Strickland and Parsons, 1972; Stainton et al., 1977; Shearer et al., 1985, among others). These primary production transfer coefficients, along with the near-surface chlorophyll concentrations, can be used to estimate phytoplankton photosynthesis. Unfortunately, if primary production transfer coefficients for a targeted water body are inaccessible, the use of surrogate values of primary production transfer coefficients from comparable waters is not as potentially reliable as is use of surrogate optical cross section spectra from comparable waters.

5. Evaluate the appropriateness of using the estimated value of phytoplankton photosynthesis thus obtained as an approximation of primary production.

This skeletal and simple listing of deterministic steps for space monitoring of inland and coastal primary production defines two separate yet interdependent and tandem experimental activities. Optical data acquired in space and *in situ* are required to obtain water-leaving spectroradiance; transference of visible radiation through the air–water interface; attenuation coefficients; and depth profiles of E_{PAR}. Water samples are simultaneously required for laboratory analyses to determine aquatic composition, from which multivariate modeling and further laboratory analyses are required to determine optical cross section spectra and primary production coefficients.

Thus, direct measurements along with multivariate optimization, atmospheric correction, and primary production models are interutilized to remotely estimate the magnitude of phytoplankton photosynthesis to be then considered as an estimate of primary production.

It is not unreasonable to anticipate that valuable inland and coastal primary production products might soon be available. Such products may not be as free of criticism as comparable products of mid-ocean primary production. Nonetheless, they will be of consequence to environmental stewardship and are eagerly awaited.

5.6   Monitoring the extent and progress of blue-green algae blooms and red tides

Once the complex and frustrating problem of obfuscation of water color by atmospheric aerosols is solved or otherwise circumvented, maps of chlorophyll concentration distribution may be readily generated from existing algorithms or methodologies for case 1 and case 2 waters. As discussed in Section 5.5, converting these chlorophyll concentrations into primary production presents additional hurdles. Nonetheless, reliable chlorophyll maps may be obtained from satellites, a reliability that, understandably, will increase as the concentration of local chlorophyll increases and the dominance of chlorophyll as a CPA increases. Both these conditions are satisfied by the presence of algal plankton blooms. It is virtually impossible to classify ocean, coastal, or inland waters into layers or regions of individual varieties of photosynthetically active chlorophyll pigmentations because mixtures of blue-green, green, red, and brown algae may be found at any depth. However, red algae are found in ocean waters but absent from inland waters.

Figure 5.7, taken from Gower et al., 2004 (their Figure 5), illustrates the SeaWiFS NASA Level 2 chlorophyll data product for 2100 UTC, September 22, 2000. Quite evident is the dramatic Pacific Ocean blue-green diatomic algal bloom extending off the western coast of Vancouver Island. This autumnal bloom is a consequence of high nutrient water masses in the buoyancy-driven Vancouver Island coastal current where relatively fresh surface water flows into the ocean from the Strait of Juan de Fuca. Chlorophyll fluorescence calculations applied to concurrent MODIS data replicated the algal bloom location and dynamic configurations evident in the SeaWiFS image. This showed that red-band fluorescence models and full-spectrum color multivariate optimization models provide a powerful means of intervalidation (in addition to being powerful individual tools for chlorophyll extraction from satellite data when circumstances dictate).

Extensive blue-green algal blooms have also been consistently observed in the western basin of Lake Erie in recent years. This is generally attributed to a variety of synergistic environmental conditions including phosphorous reduction (thereby changing the phosphorus-to-nitrogen ratio); invasion of exotic mussels (thereby increasing water clarity); changes in the local food

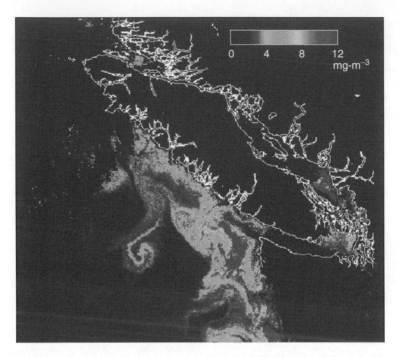

Figure 5.7 (See color insert following page 134.) SeaWiFS level 2 chlorophyll product illustrating Pacific Ocean blue-green diatomic algal bloom extending off the western coast of Vancouver Island on September 22, 2000. (From Gower, J.F.R. et al., *Can. J. Remote Sens.*, 30, 17–25, 2004.)

web; deeper penetration of visible and ultraviolet solar radiation; and general lake-wide dynamics (Charlton, 2001; Charlton et al., 2001, among others).

Ocean and inland water planktonic blooms are noticeable in concert with spring run-offs into coastal oceans, lakes, and streams. Most, if not all, of these blooms are deleterious to established dynamic equilibria of local food webs. However, in mid-ocean reaches far removed from the main influences of anthropogenic run-offs, pelagic spring planktonic blooms provide sustenance for ambient food web dynamics. The different factors responsible for declining food-fish stocks are now being intently scrutinized. In urgent cases such as Pacific Ocean salmon and Atlantic Ocean cod, in which stocks are near depletion, an ever sharpening focus on wild fisheries and aquaculture is taking place.

It has long been hypothesized (e.g., Murray and Hjort, 1912, and others) that availability of food during the critical developmental period of fish larvae determines year-classes of fish populations. In a landmark work, Platt et al. (2003) have combined satellite determinations of ocean-scale pelagic water color (from POLDER, CZCS, and SeaWiFS) with a long-term data set of haddock recruitment off the eastern continental shelf of Nova Scotia to illustrate the validity of this long-standing hypothesis. The work

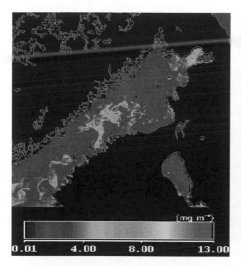

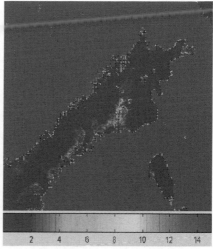

(a) Atmospheric algorithm one

(b) Atmospheric algorithm two

Figure 2.2 Aquatic chlorophyll maps acquired from August 1, 1999, SeaWiFS data over the Gulf of Finland using the same aquatic bio–geo-optical model with two different atmospheric correction algorithms. (Mapping provided courtesy of Dimitry V. Pozdnyakov, Nansen International Environmental and Remote Sensing Center.)

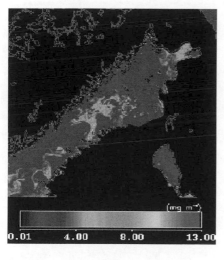
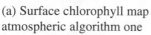

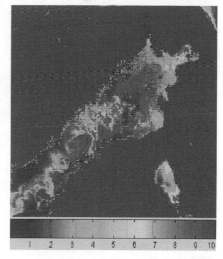

(a) Surface chlorophyll map
atmospheric algorithm one

(b) Suspended sediment map
atmospheric algorithm two

Figure 2.3 A comparison of the chlorophyll map emergent from the use of one atmospheric algorithm with the suspended sediment map arising from the use of a different atmospheric correction algorithm on the August 1, 1999, SeaWiFS image over the Gulf of Finland. The same bio–geo-optical multivariate model was used subsequent to Level 1 radiance correction by each atmospheric algorithm. (Mapping provided courtesy of Dimitry V. Pozdnyakov, Nansen International Environmental and Remote Sensing Center.)

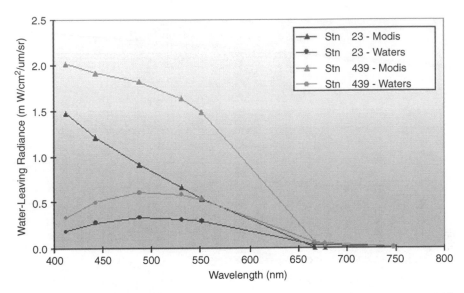

Figure 3.3 Comparison of WATERS-derived water-leaving radiance spectra with those obtained from MODIS during May, 2001. (From Bukata et al., *Backscatter*, winter issue, 29–31, 2003.)

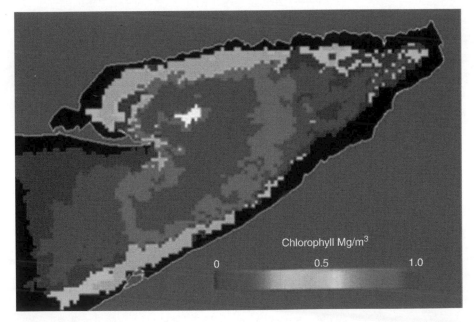

Figure 3.4 Chlorophyll concentrations in Lake Erie as determined from a simple unsupervised classification of an "equivalent WATERS" image generated from level 1 MODIS data over Lake Erie. (From Bukata et al., *Backscatter*, winter issue, 29–31, 2003.)

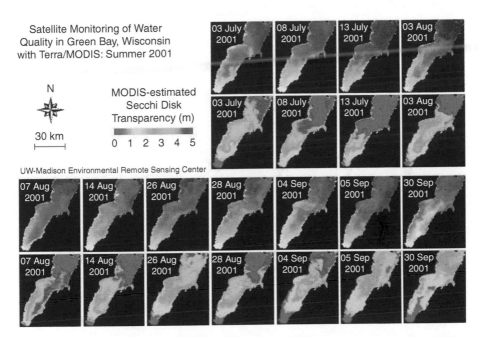

Figure 5.1 Time series of Secchi disk transparency maps generated from Terra/MODIS images of Green Bay, Wisconsin, over the summer of 2001. (From Richardson, L.L. et al., *Backscatter*, winter issue, 26–31, 2002.)

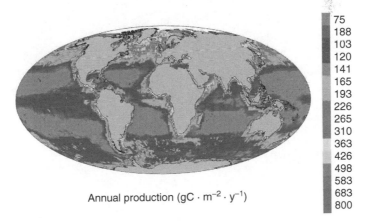

Annual production (gC · m⁻² · y⁻¹)

Figure 5.2 Annual mean field of total primary production of phytoplankton in the euphotic zone of the oceans, in g C m⁻² y⁻¹ (From Longhurst, A. et al., *J. Plankton Res.*, 17, 1245–1271, 1995).

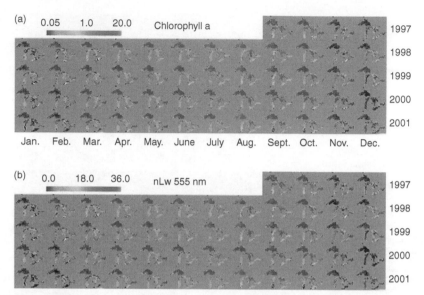

Figure 5.4 Time-series SeaWiFS data water quality features collected over the Laurentian Great Lakes: a) average chlorophyll concentration; b) average water-leaving radiance at 555 nm. (From Gower, J.F.R., *Can. J. Remote Sens.*, 26–35, 2004.)

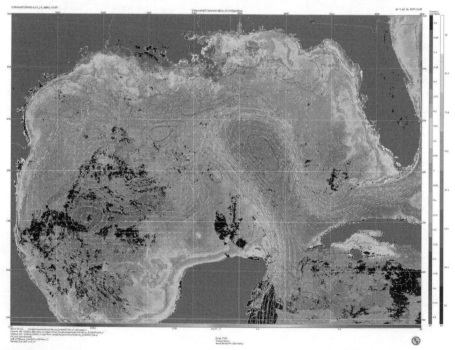

Figure 5.5 Chlorophyll concentration in the Gulf of Mexico as recorded by SeaWiFS on July 14, 2004, using the NASA OC4 chlorophyll extraction algorithm. Also shown are contoured sea surface heights and salinity. (Provided courtesy of Robert A. Arnone, Naval Research Laboratory, Stennis Space Center.)

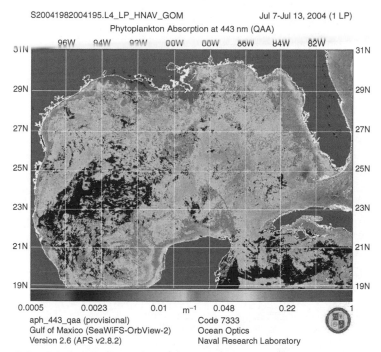

0.0005   0.0023   0.01   m⁻¹   0.048   0.22   1

aph_443_qaa (provisional)   Code 7333
Gulf of Maxico (SeaWiFS-OrbView-2)   Ocean Optics
Version 2.6 (APS v2.8.2)   Naval Research Laboratory

Figure 5.6 Spatial distribution of the phytoplanktonic absorption coefficient at 443 nm, $a_{chl}(443)$, over the Gulf of Mexico averaged over the SeaWiFS data of July 7 to 13, 2004, using an empirical quantum adiabatic algorithm. (Provided courtesy of Robert A. Arnone, Naval Research Laboratory, Stennis Space Center.)

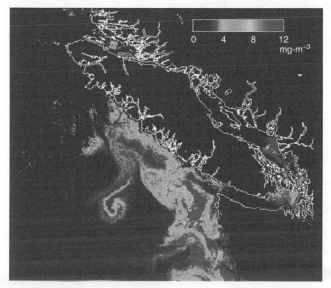

Figure 5.7 SeaWiFS level 2 chlorophyll product illustrating Pacific Ocean blue-green diatomic algal bloom extending off the western coast of Vancouver Island on September 22, 2000. (From Gower, J.F.R. et al., *Can. J. Remote Sens.*, 30, 17–25, 2004.)

Figure 5.8 Red tide on the west coast of Vancouver as recorded by an airborne CASI spectrometer in September, 1990. (Provided courtesy of G.A. Borstad Associates, Ltd. and Jim F.R. Gower, Institute of Ocean Sciences.)

Figure 5.13 Computer printout of the sum of ERTS 1 MSS bands 4, 5, and 6 as recorded over Point Pelee on March 26, 1973. Clearly delineated are the sediment transport flow around Point Pelee and the dramatic Archimedes' spiral/vortex flow pattern extending from its southeast tip. (From Bukata, R.P. et al., *Verh. Int. Verein. Limnol.*, 19, 168–178, 1975.) (http://www.schweizerbart.de).

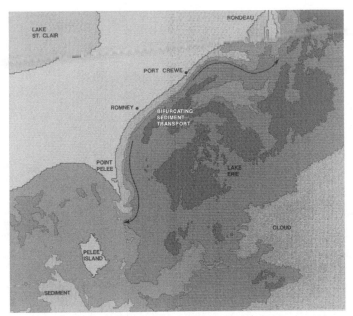

Figure 5.14 Computer printout of ERTS-1 MSS band 5 (generic aquatic turbidity map) over the western and central basins of Lake Erie on March 26, 1973, from which the details of a "mirror-image" bifurcating sediment transport model of the temporal evolution of the Point Pelee and Rondeau Landforms readily emerged. (From Bukata, R.P. et al., *Verh. Int. Verein. Limnol.*, 19, 168–178, 1975.) (http://www.schweizerbart.de).

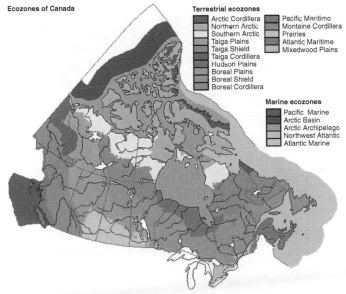

Figure 5.15 The continental and oceanic ecozones of Canada. (Public domain courtesy of Environment Canada's EcoMAP computer module.)

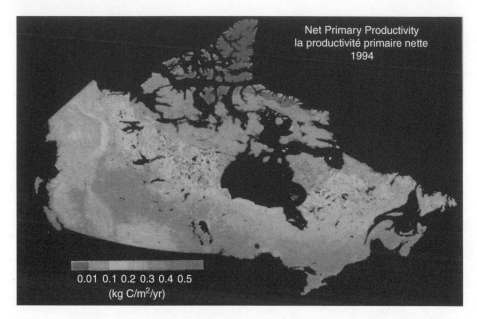

Figure 5.16 Net primary productivity (kgC m^{-2} y^{-1}) land cover map of Canada inferred from AVHRR and Landsat-7 data during 1994. This is the terrestrial equivalent to the global ocean productivity map of Figure 5.2. (Public domain courtesy of the Canada Center for Remote Sensing, CCRS, of Natural Resources Canada, NRCan).

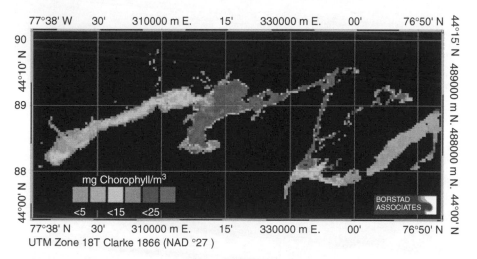

Figure 6.1 Simulated MERIS surface chlorophyll map from a CASI overflight of the Bay of Quinte (on Lake Ontario and stretching north to the Trent River Valley) on September 21, 1992. (Provided courtesy of G.A. Borstad Associates.)

convincingly illustrated that a link does exist between the timing of spring algal blooms and larval survival (the earlier the bloom, the better the survival probability).

Although an early spring phytoplankton bloom is not a guarantee of high larval survival in that year (zooplankton is the principal diet of haddock larvae), it is a recognized contributing factor to larval survival. The validity of the hypothesis, an important consideration in the analyses of dwindling food-fish stocks, could not have been accomplished without the large-scale capability of ocean color monitoring from space. After a century of ocean study restricted to using oceanographic research vessels, the hypothesis remained a hypothesis. Platt and his collaborators' work is an excellent illustration that space monitoring of aquatic resources can indeed generate a public sector product for recognizable public good.

Sidebar

Validation of the hypothesis that availability of food during the critical developmental period of fish larvae determines year-classes of fish populations is important to the global problem of diminishing fish populations. However, the most glaringly conspicuous problem remains the combination of blatant national and international overfishing and apparently ineffective enforcement policies. In this regard, an ambitious plan to reduce this overfishing also involving use of satellites has recently been proposed by Clover (2004). In addition to enforced catch quotas, fishing fleet cutbacks, and reduced food-fish consumption, he suggests that the European Union create large marine systems of fish reserves in the North Sea and the North Atlantic Ocean from which all fishing vessels would be banned. He further suggests that the public monitor adherence to this ban via free access to updated information on the Internet. This would be accomplished by legislated monitoring units mounted on fishing vessels, with their locations tracked by satellite and posted on designated Web sites. This intriguing concept of allowing citizens to police the fishing activities that are detrimental to the survival of one of their principal sources of food and economic welfare, while not without inherent difficulties, is consistent with our proposed philosophy of bringing the sponsoring public into the "inner circle" of the remote sensing community to illustrate that satellite monitoring of aquatic resources can, indeed, produce a product for common public good.

From these discussions of blue-green algae, it is evident that, in addition to obvious nuisance, algal blooms may be benign, deleterious, or beneficial, giving rise to the differentiating classification term *harmful algal blooms* (HABs). Most species of phytoplankton are not harmful and are integral to the energy transfer processes that dictate dynamics of the aquatic food web. However, some algal species produce potent neurotoxins that can invade and be transferred through the food chain, thus affecting not only the base of the food web, but also higher order shellfish, seabirds, sea mammals, and, ultimately, humans and wildlife. HABs refer to blooms of toxic or nontoxic algae that otherwise exert a negative impact on an established dynamically stable ecosystem. Algal blooms associated with anthropogenic run-offs into lakes, rivers, streams, and near-coastal oceans (i.e., into case 2 and inland case 1 waters) are invariably HABs.

A natural (as opposed to an anthropogenic) bloom is the *Red Tide* composed of fast growing oceanic red algal species that, when they accumulate into dense visible surficial patches, often impart a reddish color to the water. "Red Tide" is a misnomer, however. The blooms are not always red (aquatic discolorations may also be orange, brown, violet, or yellow) and they are not associated with ocean tidal motions. They are generally not toxic (although those that do produce toxins are a serious threat to sea life and its higher order predators). Red Tides are generally triggered by prevailing winds that cause cold nutrient-rich seawater to rise (thermal-induced process of upwelling) from deeper regions to the surface; here, through combined interactions of temperature, salinity, and solar radiation, blooms of dinoflagellate phytoplankton result.

Red Tides have been observed for a considerable time and occur throughout the globe. Perhaps the greatest media coverage has been given to the repetitive outbreaks that occur along the South African coast, although Red Tides have had a drastic impact on fishing and fisheries (usually through toxin contamination of bivalve shellfish such as oysters and clams) in Scandinavia, Japan, the Caribbean and South Pacific, and the U.S. Certainly, anthropogenic pollution injected into coastal areas has enhanced Red Tide blooms in a large part of the world. However, Red Tides that are free from toxic algae (for example, those commonly observed in the Gulf of Florida) appear to play a beneficial ecological role in the aquatic environment. The purpose, functions, and abatements of Red Tides and blue-green algal blooms have been and are being researched at numerous oceanographic institutes (see, for example, Pan et al., 1991; Sathyendranath et al., 1997; Yoo and Jeong, 1999; Charlton et al., 2001; and texts edited by Cosper et al., 1989; Okaichi et al., 1989; Graneli et al., 1990; and Smayda and Shimizu, 1993).

Figure 5.8, provided through courtesy of Gary Borstad and Jim Gower, illustrates a Red Tide on the west coast of Vancouver as recorded by an airborne CASI spectrometer in September, 1990. As mentioned, some coastal algal blooms are deleterious and can kill fish in net pens and upset the ambient aquatic food chain dynamic. Differentiation between HABs and benign blooms is not yet possible using remote sensing alone. However, used

Figure 5.8 (See color insert following page 134.) Red tide on the west coast of Vancouver as recorded by an airborne CASI spectrometer in September, 1990. (Provided courtesy of G.A. Borstad Associates, Ltd. and Jim F.R. Gower, Institute of Ocean Sciences.)

in conjunction with ancillary data and prior experience, space observations of water quality might soon assist in the identification of high-risk blooms.

5.7 Delineating the presence of inorganic turbidity in inland and coastal waters

Of all the applications of space-acquired data to inland and coastal water quality monitoring, delineating the presence of inorganic particulates (not necessarily the concentrations) is undoubtedly the simplest and most convincing. However, for waters in which suspended sediments are known to be overwhelmingly dominant CPAs, delineating concentrations is also fairly straightforward and convincing.

Although diverse geology produces diverse species of minerals, suspended minerals in heavily sediment-laden inland waters are to limnological remote sensing algorithms as chlorophyll (despite the diversity of ocean algal species) is to oceanographic remote sensing algorithms. In each case, the optical modeling and multivariate optimization methodology reduces to consideration of molecular water plus a single dominant CPA (optically binary waters). Delineating extent and progress of chlorophyll is important to estimating aquatic bioproduction in addition to providing temporal and spatial information on ocean currents and ocean dynamics; therefore, having chlorophyll as that single CPA is a major advantage.

Nonetheless, use of inorganic turbidity to provide information on lake currents and lake and river dynamics should not be underestimated. Suspended sediments carried by water masses controlled by currents and sea states generated by winds, temperatures, and other meteorological forcing functions afford excellent opportunities to monitor and understand limnological phenomena such as

- Littoral drift
- Upwelling and downwelling episodes
- Inversion of lake-wide thermal regimes (stratification, the seasonal evolution of a warmer upper layer, *epilimnion*, distinct from a cooler lower layer, *hypolimnion*, and separated by a *thermocline*)
- Shore erosion
- Shore accretion
- Temporal evolution of distinctive shore features such as spits, harbors, isthmuses, shoals, bays, and others

Shortly after the launch of ERTS-1, space monitoring of water-borne suspended inorganic matter showed how littoral drift and geophysical processes were integral to development of a "mirror-image" bifurcated sediment transport model for the origin and temporal evolution of the Point Pelee and Rondeau landforms on the north shore of Lake Erie. (See Figure 5.11 through Figure 5.13 later in this chapter.)

Water quality (drinking, recreational, farming) concerns require reliable *in situ* ground-based determinations of water quality parameters (toxic and benign). Such parameter values are of consequence to the local acceptance of water quality and as meaningful inputs to predictive environmental models. Therefore, it is essential that *in situ* measurements of a water quality parameter be intercomparable at each station as well as among stations. This requires that monitoring systems be intercalibrated as well as the water masses that are being monitored (especially between stations). A string of water quality monitoring stations along a long transport avenue or evacuation route, such as the Mississippi, St. Lawrence, Amazon, Nile, or other major global river, often does not sample water masses associated with a common transport zone.

A recurrent theme in this book has been the incorporation of space monitoring and conventional ground-based monitoring. Monitoring littoral drift and pelagic sediment transport provides valuable support to ground-based monitoring networks. Synoptic overviews of current flow can

Figure 5.9 True color composite of the blue, green, and red spectral bands of the Landsat 5 TM obtained over the Lac Saint Pierre region of the St. Lawrence River on August 20, 1984. (From Bruton, J.E. et al., *Water Poll. Res. J. Can.*, 23, 243–252, 1988.)

assist networks in establishing meaningful sampling strategies. Figure 5.9, taken from Bruton et al. (1988), illustrates a true color composite of the blue, green, and red spectral bands of the Landsat 5 TM obtained over the Lac Saint Pierre region of the St. Lawrence River on August 20, 1984. An obvious feature of such a synoptic overview is the river-to-lake-to-river pattern of persistent, mutually independent, and spatially extensive suspended inorganic turbidity zones (consistent with the work of Duane, 1967; Bukata et al., 1974; Lachance et al., 1979).

Utilizing TM data from 15 such Landsat 4 and Landsat 5 overviews, Bruton and his collaborators showed the following:

- Persistent, mutually independent turbidity zones were extensive in space and time.
- Seasonally cyclic succession of patterns of these well-defined turbidity zones was in evidence.
- Although the turbidity zones maintained their independence, they did not maintain constant widths, constant distance from shore, or constant sediment loadings.

Visible and thermal Landsat data were used to illustrate close relationships among the distinct zonal synoptic patterns; the bathymetry of the lake and river; guiding-center zonal flow; lateral diffusion; and the near-surface aquatic temperatures.

The aquatic transport corridor extending from the Upper Great Lakes through Lake St. Clair and the lower Great Lakes and eventually evacuating into the Atlantic Ocean through the St. Lawrence River is vulnerable to uses and abuses of the human and wildlife populations that it sustains. Monitoring the shape and extents of such dynamic flow regimes from space is critical to determining the fates of injected toxic contaminants and their impacts on the environment. Clearly, improper placement of in-water stations (stations that are not within the same zones) within otherwise well-defined sampling strategies can lead to erroneous intercomparisons of inland water quality parameters (El-Shaarawi, 1984). Extended regions of temporally and spatially persistent turbidity zones are common features of basins and watersheds on a global basis. Therefore, awareness of their ephemeral behavior through the use of synoptic space overviews should play a prominent role in optimizing reliability of ground-based sampling strategies.

5.8   Monitoring the extent and progress of marine and inland oil spills

Of the variety of ways in which developed nations' use of energy can pollute the Earth's environment, the most vivid images (of petroleum-soaked sea birds, for example) in people's minds are planted by oil spills. From the extensive and intense media coverage, these spills seem to occur on a regular basis. Oil spills can and do destroy marine life as well as inflict damage upon the habitats of humans and animals. The majority of marine oil spills result from accidental rupturing of oil containers on massive ocean tankers or intentional emptying of bilge tanks prior to a ship's entry into port. Therefore, routine surveillance of shipping routes and coastal marine areas is invoked to enforce pollution laws and to identify polluters.

Oil spills are invariably spectacular in size and environmental impact; however, they are neither the sole nor the major source of oil pollution in ocean waters. Other anthropogenic sources include off-shore drilling; oil consumption in industry and automobiles; atmospheric fallout; routine ship maintenance; casual dumping of oil products down storm drains after oil changes; and urban street run-off, among others. Natural sources of marine oil pollution include seepage off the ocean floor and erosion of sedimentary rocks. Although it is difficult to ascertain precise values of the year-by-year amount of pollutant oils injected into the global oceans, it is generally and probably reasonably estimated as several hundred millions of gallons (which could make it close to about one-half the world supply). Oil spills in inland waters and waterways, although always a possible threat, are considerably less frequent than in marine waters. Oil contamination in inland water networks is predominantly the consequence of the anthropogenic uses of the populated watersheds (public and industrial).

Injection of massive amounts of oil from a ship oil spill into a dynamic open ocean or coastal environment has a negative impact upon vast areas of water and adjacent land. Spills require action by emergency response and evaluation teams ready to implement remediation measures to minimize this environmental impact. In turn, this requires identification of the spill location, size and physical extent of the spill, direction and magnitude of oil movement, and atmospheric and sea-state conditions. Remote sensing should therefore play an important role in spill situations.

Although remote sensing activities (obtaining satellite imagery and/or initiating aircraft overflying of the spill location) can be mobilized for oil spills (marine, coastal, or inland), the techniques for remotely monitoring oil spills are also appropriate for monitoring the natural and anthropogenic oil pollution of inland or ocean waters. However, in the absence of obvious oil slicks, such monitoring becomes much more difficult. This is another reason why remote monitoring of aquatic oil pollution (mainly oil spills) is and perhaps should be regarded as a higher priority for oceans than it is for inland waters and waterways. Obviously, important issues associated with oil spills include

- Tracing the spill to a responsible ship or oil drilling platform (stationary off-shore oil rigs are relatively easy to target, moving oil tankers are less easy)
- Detecting it early enough and monitoring it often enough to enable clean-up operations to be as rapid and as effective as possible
- Obtaining the identity and estimating the amount of oil product included within the spill
- Enforcing laws and penalties associated with polluting the natural environment

A variety of passive and active remote sensing devices is used to monitor aquatic oil pollution. The easiest, most frequently used, least expensive, and least numerically accurate devices are passive visible, infrared, and infrared/ultraviolet optical cameras flown from airborne platforms. Although they are of value in locating an on-water oil slick and providing reasonable estimates of its physical size and shape, such cameras are of no value in discriminating oil on beaches or on coastal vegetation. However, most off-shore oil spill surveillance is currently performed with simple still or video photography. Visible wavelength cameras or radiometers (such as those mounted on the ERTS/Landsat vehicles) are of little use beyond identifying sites and extents of offshore oil spills.

Oil displays no visible spectral signature in the visible wavelength region upon which unambiguous discrimination can be made. Instead, oil displays essentially spectrally invariant radiometric return ("slicks" or "sheens") across the visible region that is of higher surface reflectance than the water upon which it is buoyed. Thus, recording of such brighter slicks in a Landsat image can serve as an indication of possible oil spillage. Such slicks appear only slightly less "black" in visible imagery than marine water, however. Consequently, they can be mistaken for cloud shadows, sun glint, wind sheens, debris, or bottom features, and false alarms can and do frequently occur.

Commercially available thermal infrared devices (wavelength region 8 to 14 μm) possess the capability to differentiate, in a very approximate manner, relative (not absolute) oil thickness. Oil is optically thick and therefore absorbs solar radiation and re-emits a portion of this radiation as heat. Thus, in a thermal image, thick oil would appear hot, intermediate oil cool, and thin oil undetectable. However, sensors that will provide absolute measurements of oil slick thickness are not currently available. Remote sensing might play important roles in locating oil spills, monitoring their progress and physical extents, and possibly targeting the source of the oil infraction. However, lack of sensors to estimate precisely the total volume of leaked or deliberately injected oil makes court prosecution of targeted polluters difficult despite clearly evident and documented environmental damage resulting from the spill. Science and technology are currently far from solving the complex problem of estimating oil spillage volume.

Laser fluorosensors are active devices that first induce and then monitor stimulated fluorescence from in-water or on-water petroleum compounds (absorption of ultraviolet radiation followed by emission of visible wavelength radiation). These are the best currently available remote sensors for oil spill detection. Oil is a relatively unique petroleum compound that displays stimulated UV-induced fluorescence, so UV-induced fluorescence is a good indication of the presence of oil. The fluorescent response of crude oil covers the entire visible wavelength band with a moderately broad peak at ~480 nm. Recall from our discussions of trans-spectral processes in Section 3.5, that chlorophyll and dissolved organic matter are CPAs that display natural solar-induced fluorescence and that the solar-induced fluorescence of chlorophyll is recorded as a well-defined Gaussian peak in the red region (685 nm) of the upwelling visible spectrum. However, the fluorescence of dissolved organic matter is an ill-defined broadband almost full-spectrum distribution with a maximum in the range 490 to 530 nm.

Laser fluorosensors of high spectral sensitivity are capable of detecting oil from airborne and ship-borne platforms — despite the variety of fluorescence spectra and spectral peaks from naturally solar-stimulated chlorophyll and dissolved organic matter and artificially UV-stimulated crude oil, as well as largely unknown precise values of pertinent fluorescence quantum yields. They are also very effective in identifying some oil products and locating oil on sandy shores, ice, and snow. However, they require cloud-free conditions to detect oil spills. Therefore, radar, which can penetrate clouds, becomes the preferred frequency for active devices mounted on satellite platforms.

Synthetic aperture radar (SAR) sensors transmit microwave radiation in pulses and record the radiation backscattered into the sensor. SAR is sensitive to short Bragg-scale waves (centimeter-scale sea surface roughness waves generated by the near-surface wind field) independent of time of day and cloud conditions. Presence of oil on water reduces the backscatter over the area of the spill due to suppression of localized Bragg-scale waves. As a consequence, oil spills in a SAR image appear as circular or curvilinear features darker than the water that is buoying them. This is opposite to the situation for passive satellite sensors such as the Landsat MSS and TM in which oils spills appear less dark than their surrounding waters. However, as the oil spillage begins to emulsify and clean-up efforts become effective, the Bragg waves are not as effectively damped and the oil appears increasingly brighter to SAR. Here, as is the case for nonradar sensors, wind shadows near land, regions of low wind speeds, grease, and other natural surficial features may be mistakenly interpreted as oil spills.

For detailed analyses of SAR from space see Sikora et al. (1995) and Mourad (1999), among others. The first nonmilitary SAR was launched in 1978 aboard the ill-fated satellite Seasat, which was only operational for less than 4 months. A few SARS were tested during shuttle launches during the 1980s. However, several SAR launches occurred during the 1990s; two of the most recent are Canada's Radarsat-1 and Europe's ENVISAT, which features an advanced synthetic aperture radar (ASAR) sensor. Although the presence of polluting petroleum products in natural waters does affect water quality and satellite SAR images of oil slicks technically qualify as water *quality* products, such products do very little in advancing the cause of water quality monitoring from space. Radar is far more effective for monitoring water *quantity*. Water color is the recommended method for monitoring water quality.

Due to the sensitive national and international political implications of oil spills, we will refrain from illustrating actual remote-sensing images of noncontrolled oil spillage into the marine or inland aquatic environment. Suffice to say that such illustrative examples (e.g., the 1989 *Exxon Valdez* spill and the 1996 *Sea Empress* spill) do exist within the published literature, media coverage stories, and specific sites on the Worldwide Web. Simple search-engine entries should readily yield them.

5.9   Delineating the regional groundwater discharge, recharge, and transition areas (and temporal changes therein)

Delineating the location and movement of groundwater in watersheds and basins is of recognized importance to such water quantity-focused resource areas as land use and pollutant transport. Groundwater is a component of the hydrological cycle that is pivotal to hydrogeological modeling. In addition to knowledge of groundwater presence, movement, and flushing times, rigorous hydrogeological modeling requires precise incorporation of interactions among meteorological conditions, soil dynamics, basin morphology, indigenous verdure, and long-term climate. Although space observations cannot provide such precision, they can offer valuable information on groundwater flow pathways by identifying the principal regional groundwater regimes within a freshwater basin. These principal regimes are

- *Discharge areas*: regions of the basin for which the water table is in close proximity to the terrain surface; the ultimate discharge area is standing water
- *Recharge areas*: regions of the basin for which the water table is substantially distant from the terrain surface
- *Transition areas*: regions of the basin that, due to an intermediate distance of the water table from the terrain surface, are distinct from discharge and recharge areas at any instant of time, but do not display the time constancy generally associated with those areas

Transition areas display volatility to changing conditions of local weather. Under conditions of excessive precipitation, they may assume characteristics of discharge areas. During excessive drought, they may assume characteristics of recharge areas. Although transition areas are most vulnerable to floods and droughts, from an agricultural standpoint, they generally represent the most desirable farmland in a freshwater basin.

Early studies of Landsat data over the Big Creek and Big Otter Creek freshwater basins of southern Ontario illustrated that visible red and near-infrared radiances could convincingly classify the discharge, recharge, and transition areas on the basis of the water table depth, thus providing an estimate of the gradient of chlorophyll concentration in the phreatophytic (water-loving) natural and cultivated vegetative canopy over the basin (Bukata et al., 1978). Scatter-plots of L(red) vs. L(infrared) depicting the "lushness parabolas" of Bukata et al. (1978) were fore-runners of the routinely used normalized difference vegetation indices (NDVI) discussed in Chapter Four. Because groundwater generally flows from recharge to discharge through transition areas, once space data can delineate the location of these areas, the groundwater flow pathways may be confidently inferred. Figure 5.10, adapted from Figure 19 of Bukata et al., 1978, illustrates how such groundwater flow pathways (arrows) may be simply mapped (1974 Landsat data over the Scotland area of the Big Creek, Big Otter basins).

However, such groundwater movements delineated from satellite altitudes do not give any direct information on the quality or quantity of water within the labyrinth of transport avenues. Whiting (1976, 1984) has conducted work on the considerably more brackish waters of the Big Quill Lake and Little Quill Lake watersheds in Saskatchewan. This work illustrated emergence of complex combinations of upright and inverted "lushness parabolas" that depended upon the interplay among groundwater flow, basin vegetation that flourished under freshwater, and basin vegetation that flourished under saline water.

During the dramatic *glasnost* and *perestroika* years of political transformation, the (then) Soviet Academy of Science adopted the "lushness parabola" concept (see the updated revisitation of the work in Bukata et al., 1991c). A prevalent theory as to the devastating Chernobyl nuclear power station disaster is that the Chernobyl site did not possess suitable drainage avenues for its heavy water coolant. The "lushness parabola" concept (now

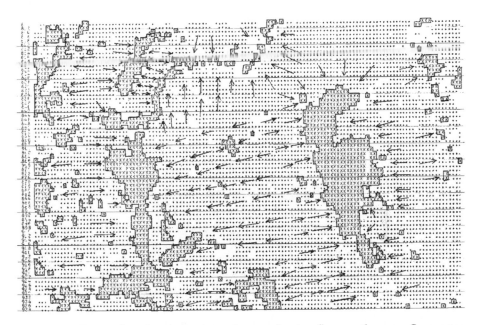

Figure 5.10 First-order determination of groundwater flow pathways. Computer printouts of correlated visible and infrared radiances recorded by Landsat 1 MSS on July 6, 1974, over the freshwater Big Creek and Big Otter Basins in southern Ontario were used to distinguish discharge, recharge, and transition regions according to "lushness" of the vegetation supported by proximity of the water table to the surface. The arrows delineate groundwater movement from recharge through transition through discharge areas. (From Bukata, R.P. et al., *Can. J. Spectrosc.*, 23, 79–91, 1978.)

computerized NDVI modules) is being used to analyze the appropriateness of future proposed sites of Russian reactors in hopes of preventing nuclear disasters. Delineation of groundwater flow pathways from space could therefore act in a similar capacity in countries that are considering nuclear power as an alternate or increasingly required additional energy source due to escalating demands for energy.

This application of remote sensing to inland groundwater monitoring contributes no information to basin water quality or basin water quantity. Nevertheless, it does provide information as to the presence, possible presence, and directions of subsurface flow of water within the basin. Such information is invaluable to the farming, recreational, commercial, health, and energy concerns of developed and developing countries. However, when lushness parabola or NDVI analyses are applied to *standing water*, as opposed to terrestrial targets (and therefore groundwater), estimates of quality of inland (or coastal) water and quantity of aquatic components are possible. The converse responses of chlorophyll to visible red wavelengths (strong absorption) and infrared wavelengths (strong reflectance) are integral to discussions of the red tide and blue-green algal blooms in Section 5.7.

5.10   Monitoring inland and coastal water quantity

Over the years, hydrology (the science that deals with the water cycle) has been better served by remote sensing than has limnology (the science that deals with the life cycle of standing inland water). This is due, in no small measure, to water cycle processes being defined by water *quantity* parameters as opposed to water *quality* parameters. Water quantity parameters include surface area, water level, water flow diversions, ice onset and break-up, snow cover, permafrost, soil moisture, and other physical variables that may be directly measured or inferred from other physical variables.

The processes involved in such inference are also physical in nature and deal with change of state (freezing, evaporation, transpiration, and sublimation), rates of change of state (time derivatives of freezing, evaporation, transpiration, and sublimation), and meteorology (temperature, wind, atmospheric composition, vapor pressure, and dew point). Thus, because it is largely a single disciplinary (physics) venture, hydrology parallels space science in the development and applications of models, algorithms, and space-derived aquatic products. However, as for limnology, biological complexities emerge when water quality and water quantity are collectively considered as interactive determinants of environmental ecosystem health.

General circulation models (GCMs) are generally accepted tools for studies of potential global change (principally potential global climate change) because they are global in scale and based on the laws of physics. However, several competing GCMs based on physical laws are not always based on the *same* physical laws. Furthermore, even if all GCMs were based on the same physical laws, there would still be problems with their value when used as *predictive models*, particularly when used as *predictive impact models* (invariably in a "what if?" mode). These problems in prediction of environmental impacts are consequences of two basic departures from dispassionate scientific logic that arise from:

- Reliance upon subjective assumed relationships that an increase or a decrease over time of a particular environmental variable is good or bad for another environmental variable. Some relationships, although difficult to formulate mathematically, are nonetheless intuitive (e.g., if the sun burns out, chances are that the Earth will get cold and a lot of life will terminate). However, most relationships, unless sufficient scientifically defendable data exist to establish them irrefutably, are not intuitive; intuition invariably yields to opinion and opinion becomes accepted as fact.
- Extrapolation of a current value of an environmental variable 10, 20, or 30 years into the future (based on worst-case "what if?" scenarios) and insertion of those values dutifully into the physics-based GCM to predict a future global climate. Then, using some previously derived "intuition turned opinion turned fact" relationship, the GCM is used as supportive evidence of a future environmental impact

(generally a dire impact because "dire" attracts more media headlines and funding-agency priorities than does "not dire").

These and other problems associated with GCMs notwithstanding, hydrologic and atmospheric variables are mainstays of such physics-based climate change models. Remote sensing, therefore, affords excellent opportunities to provide terrestrial and aquatic inputs to GCMs. As such, however, remote sensing must be realistic and must safeguard itself from becoming unwittingly used to provide false scientific legitimacy to global-change spin-doctoring from prevalent (and invariably transient and conflicting) political, social, media, theological, financial, and/or academic ideologies. This warning will be reiterated in Chapter Seven.

The water cycle is driven by dynamical equilibria established by laws of physics. The principal and directly measurable variables of water cycle dynamics are water state and water quantity. Apart from injected pollutants (from earthquakes, volcano eruptions, forest fires) that comprise a water *quality* threat and require disaster management protocols based on local mitigation studies, most natural (as opposed to anthropogenic) disaster threats to natural waters are *quantity* issues. Thus, from a space monitoring perspective, disasters inflicted by hurricanes, tidal waves, tsunamis (large sea-wave responses to marine disturbances), advance and/or retreat of polar ice fields, glacier and ice-floe migration, El Niño events, waterspouts, and other physics-driven phenomena become issues of near real-time detection, location, progression, and estimate of water quantity.

Unlike determination of water *quality*, which demands full-spectral digital data formats and bio–geo-optical models and algorithms that are invariably local in application, determination of water *quantity* capitalizes upon single and broadband wavelengths (e.g., SAR, panchromatic, all-weather) and the use of remote sensing data in an aerial photographic image format. As a remote sensing tool, photogrammetry is not as locally restrictive as bio–geo-optical modeling. Thus, an enviable advantage is provided to determination of water quantity from space; this explains, in part, why remote sensing has better served the interests of hydrology than the interests of limnology.

Photogrammetry is also conducive to the use of remote sensing with geographic information systems (GISs). A GIS is a vital tool for making use of high-resolution visible and SAR data for aquatic disaster mitigation by extracting topography (through use of available digital elevation model, DEM, computer programs) and preparing land-use maps of the affected or threatened basin, watershed, or coast. Rapidly developing Internet Web technology has enabled scientists, computer experts, government institutes, and educators to establish Web sites devoted to weather forecasting and illustrations of natural disasters as support tools for decision makers and thereby assist in the implementation of disaster management protocols. The Internet is rife with satellite observations of hurricanes, ENSO events, glaciers, polar ice caps, and other marine weather-related and climatologic features. Simple search engine commands will readily link to them.

Perhaps the most important and most commonly space-monitored inland water quantity natural disaster is flooding of lakes and rivers. Floods are to continental waters as hurricanes and tsunamis are to coastal ocean waters. Conventionally, flood mapping is accomplished by the overlaying of a preflood remote sensing image with a peak-flood image to delineate the inundated area.

Aside

This common practice of delineating image overlays to determine the temporal evolution of affected floodplains is an excellent example of our advocated compromise between defendable science and defendable use of that science. Ironically (and to the embarrassment of the author), it also serves as an admitted example of a lost opportunity to convey the value of a similar remote sensing water product delineating the areal extents of wetlands affected by persistent changes in standing water levels (see Section 5.14).

Every major river on Earth has suffered the ravages of the seasonal vagaries of local climate (rapid snow melt, heavy precipitation, ice-jamming, etc.). From space and ground observations, damages are assessed in terms of property, crops, and environmental impact. Remote sensing and GIS methodology are valuable assets to the development of emergency measures and disaster management practices as well as to the placement of ground-based infrastructure to avert or otherwise abate or benefit from future recurrence of high waters. Again, the Internet contains numerous illustrations of flooding occurring on the Mississippi, Nile, Thames, Yangtze, Amazon, Po, and other major rivers.

Detailed information on flood monitoring in particular and hydrological mapping from space in general may be found in Barrett et al., 1990; Whiting and Bukata, 1990, 1995; Barber et al., 1996; Brakenridge et al., 1998; Townsend and Walsh, 1998; Simonovic, 1999; Sanders and Tabuchi, 2000; Sugumaran et al., 2000; and elsewhere.

In 1997 the National Water Research Institute's POES/AVHRR tracking station monitored the floodwaters of the north-flowing Red River in Manitoba, Canada. Figure 5.11 displays one of the NWRI AVHRR images acquired on April 27, 1997, when the flood crest was at Emerson, North Dakota. Two weeks after the flood crest had passed Emerson, large areas remained under flood conditions and the Red River continued to flow at flood stage. On the image are sketched the 1997 flood area and, for comparison, the 1979 flooded area. The city of Winnipeg, however, was virtually unscathed by the major flooding due to construction of the Red River floodway in the early 1960s. This saved inestimable damage and cost to the city and Manitoba. (Note

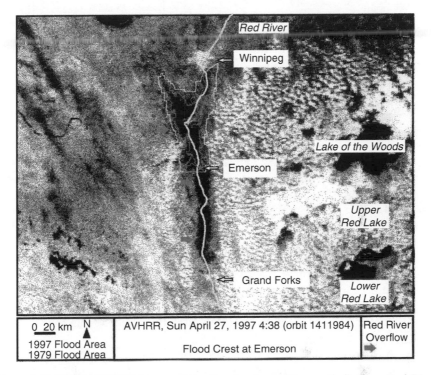

Figure 5.11 POES/AVHRR image of the floodwaters of the north-flowing Red River in Manitoba, Canada, acquired by the National Water Research Institute's satellite tracking station on April 27, 1997.

the similarities between the approaches that resulted in Figure 5.11 and Figure 5.17.)

5.11 Recording natural and anthropogenic changes in shoreline/coastal features

The geophysical morphology of watershed and continental shorelines is influenced by numerous dynamic processes resulting from air–water, air–land, and land–water interactions. Techniques available for monitoring and analyzing such phenomena (Haras et al., 1976, among others) vary from historical records dating back to early exploration days (survey records, photographs, navigational charts, and anecdotes) to ground-based measurements at a specific site (shore profile networks, current meters, ground control points, and aerial photography) to synoptic overviews of entire basins from satellite altitudes (photogrammetry, current flow models, water quantity models, and classification methodology). Shoreline changes are generally recorded as erosion (regression), accretion (advance), and flooding (inundation). Whether shorelines are altered by natural or anthropogenic forcing functions, the rates at which these three shoreline responses occur are directly controlled by quiescent and catastrophic events associated with local climatic and meteorological conditions. Clearly, shoreline changes have a direct impact on the health and habitat of human and wildlife dependent upon the watershed or coastal region for sustenance.

In addition to monitoring the changes in shoreline evolution, water quantity, and standing water quality, remote sensing affords excellent opportunities for value-added information on the geophysical processes responsible for the natural evolution of shoreline land-forms. To exemplify the role of synoptic space-acquired data in illustrating the origin of conspicuous shoreline features, as well as that such a role for space data was quickly recognized, we shall revisit some very early ERTS-1 (prior to its name change to Landsat-1) work of Bukata et al. (1975). ERTS-1 MSS data acquired over western Lake Erie on March 26, 1973, were used in conjunction with aircraft-acquired vertical photography over Point Pelee (a knife-edge landform on the northwestern shoreline of Lake Erie with relict beach ridges on its western side) on April 14, 1973. Taken from Figure 3 of Bukata et al. (1975), Figure 5.12 illustrates the April 14 vertical aerial photograph (from 2740 m) of Point Pelee. Obvious features of the off-shore turbidity map include

- A distinct swirl pattern emanating southeastward from the southern tip of the landform
- A suspended sediment load considerably more pronounced on the eastern shoreline than on the western shoreline of Point Pelee
- A circular vortex structure with a series of Archimedes' spiral flow lines (expanding radial locus r defined as $r = k\theta$, where θ continually cycles from 0 to 360°) encompassing the vortex core and linking the expanding vortex structure to the tip of Point Pelee

Figure 5.13 (Figure 4 from Bukata et al., 1975) illustrates one of the first Canadian attempts at displaying data from a computer-compatible ERTS tape on a cumbersome mainframe system with paper readout — a crude computer printout of the sum of bands 4, 5, and 6 MSS channels as recorded over Point Pelee earlier on March 26. The fluid mechanical flow around Point Pelee and the delineation of the Archimedes' spiral/vortex pattern are indeed striking (although departure from geographic fidelity is a consequence of computer printout line spacing).

Figure 5.14 is another crude computer printout (taken from Figure 2 of Bukata et al., 1975) of the ERTS-1 MSS band 5 data (general aquatic turbidity map) over the western and central basins of Lake Erie on March 26, 1973. High concentrations of suspended sediment are evident in the shallow western basin. Coastal sediment transport is evident along the north shore of the central basin between the distinct landforms Point Pelee (knife-edge shape) on the west and Rondeau (considerably less than a knife-edge shape). The most significant feature of Figure 5.14, however, is the indication of bifurcating sediment transport along the north shore between Point Pelee and Rondeau with an apparent stagnation point somewhere between Romney

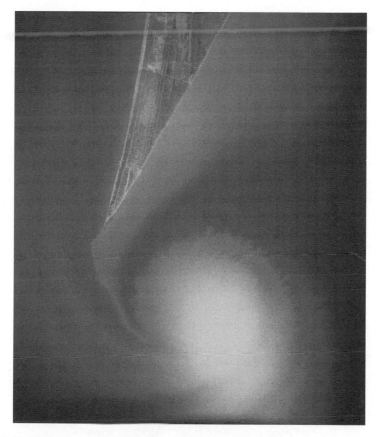

Figure 5.12 Vertical aerial photograph of Point Pelee (landform on the northern shore of Lake Erie) taken on April 14, 1973. (From Bukata, R.P. et al., *Verh. Int. Verein. Limnol.*, 19, 168–178, 1975.) (http://www.schweizerbart.de)

and Port Crewe. From this stagnation point, sediment is transported westward around the tip of Point Pelee and eastward around Rondeau.

A closer look at the pixels defining the Rondeau area indicated a geophysical formation on its west side similar to but considerably less distinctive than the Archimedes' spiral configuration on the east side of Point Pelee. Using eight such synoptic satellite overviews (March to October, 1973) along with *in situ* current meter data, a "mirror-image" model based upon bifurcated current transport of suspended sediment was devised to successfully explain

- The natural evolution of the Point Pelee and Rondeau landforms (including the knife-edge shape of Point Pelee and the less dramatic shape of Rondeau)
- The ubiquitous presence of geophysical aquatic turbidity features to the east of Point Pelee and to the west of Rondeau

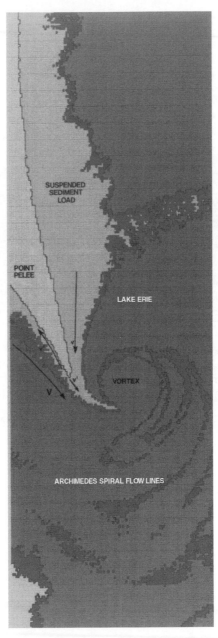

Figure 5.13 (See color insert following page 134.) Computer printout of the sum of ERTS 1 MSS bands 4, 5, and 6 as recorded over Point Pelee on March 26, 1973. Clearly delineated are the sediment transport flow around Point Pelee and the dramatic Archimedes' spiral/vortex flow pattern extending from its southeast tip. (From Buka-ta, R.P. et al., *Verh. Int. Verein. Limnol.*, 19, 168–178, 1975.) (http://www.schweizer-bart.de).

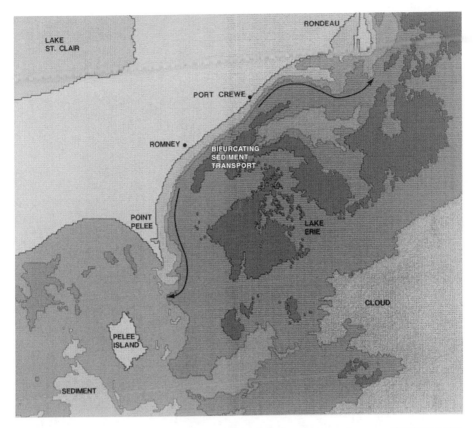

Figure 5.14 (See color insert following page 134.) Computer printout of ERTS-1 MSS band 5 (generic aquatic turbidity map) over the western and central basins of Lake Erie on March 26, 1973, from which the details of a "mirror-image" bifurcating sediment transport model of the temporal evolution of the Point Pelee and Rondeau Landforms readily emerged. (From Bukata, R.P. et al., *Verh. Int. Verein. Limnol.*, 19, 168–178, 1975.) (http://www.schweizerbart.de).

- The formations of relict beach ridges on the west side of Point Pelee and the east side of Rondeau

Archimedes' spirals, geophysical phenomena, sediment transport, bifurcated and winding currents, and other events and processes occurring along shores and coastlines — particularly if displayed in a continual or continuous pictorial time-series format — become an attention-grabbing data product that may be readily presented and explained to an end-usership in a manner akin to the manner in which the serpentine meanderings of the jet stream are explained to the end-usership of television weather channels. The tremendous technological advances (since the early days of remote sensing from space) in high speed computerized data handling and analyses systems coupled with much easier access to digital data provide exciting opportunities to utilize space observations of coastal regions to establish and/or validate models of water transport phenomena, coastal evolution, and land-water interactions, as well as to illustrate their value to environmental stewards and policy-makers.

5.12 Recording changing configurations of continental ecozones

In Section 4.2.1 we introduced the terms *ecozones* (ecologically distinct areas of the Earth's biosphere that have assumed permanently recognizable features such as trees, wetlands, mountains, and plains) and *ecotones* (transition areas between adjacent ecozones that display characteristics, including plant and animal species, of each ecozone). As an illustration of continental and ocean coastal ecozones, Figure 5.15 (GIS-generated EcoMAP provided by Environment Canada) displays the continental and coastal ecozones of Canada.

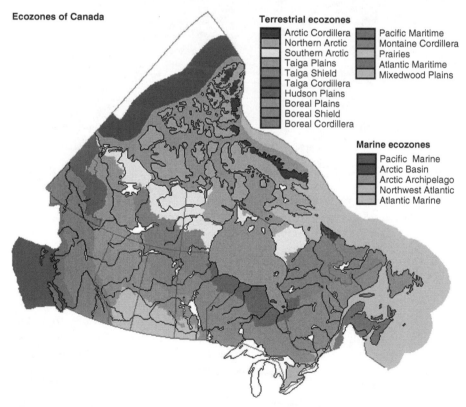

Ecozones of Canada

Terrestrial ecozones

Arctic Cordillera
Northern Arctic
Southern Arctic
Taiga Plains
Taiga Shield
Taiga Cordillera
Hudson Plains
Boreal Plains
Boreal Shield
Boreal Cordillera

Pacific Maritime
Montaine Cordillera
Prairies
Atlantic Maritime
Mixedwood Plains

Marine ecozones

Pacific Marine
Arctic Basin
Arctic Archipelago
Northwest Atlantic
Atlantic Marine

Figure 5.15 (See color insert following page 134.) The continental and oceanic ecozones of Canada. (Public domain courtesy of Environment Canada's EcoMAP computer module.)

Ecozone mapping is the manner by which the biodiversity of a country or region is commonly defined. It is based upon the compilation of a national hierarchy of ecological units that systematically divide the country or region into smaller and smaller areas of similar physical and biological characteristics. Characteristics such as climate, geology, soils, hydrology, and communities of indigenous species of plants, animals, and microorganisms are considered, as are the synergistic environmental consequences of local interplay of social, cultural, economic, and other anthropogenic activities within the region. Local synergy renders the use of an ecozone mapping module nonuniversal in space or time. Among natural, anthropogenic, and responsive activities and processes, it also plays a major role in the location, characteristics, and temporal evolution of the country's or region's ecotones. Thus, Figure 5.15 is merely the most current map of Canada's ecozones.

As the world's second largest country, Canada's geographical expanse ensures containment within its sovereignty of regional scale regimes of differing tectonic and climatic conditions. As seen from Figure 5.15, Canada comprises 15 terrestrial ecozones (ranging from the Arctic cordillera through the taiga, Hudson, and boreal plains, the prairies, and on to the Atlantic maritime) and five coastal marine ecozones (displayed in the 200-nautical mile exclusive economic zone, EEZ). The linkages running through the labyrinth of ecozones are, of course, the ecotones, which, understandably, are much more nebulously defined. Like other industrialized nations, Canada has evolved through highly correlated relationships between society and climate. Canada also has a resource-based economy with its peoples living and working in a variety of climatic regimes (southern prairies, valleys within the cordillera, coastal regions of the Laurentian Great Lakes, native settlements in the northern limits of tundra and subarctic woodlands, and sparse habitation in boreal and taiga forests, permafrost regions, Arctic ice fields, and lake and riverine wetlands and grasslands).

Such ecozone maps and the time snapshot delineation of the changing features of a country's landscape that they represent serve to illustrate ecosystem response to cumulative environmental stress and to provide a possible early warning of impending environmental risk. More disturbingly, however, such maps address the lack of opportunity for utilizing existing time-series of satellite data as a means of monitoring and assessing environmental behavior. A great archive exists that is made up of visible and thermal repetitive coverage of most of the Canadian (and American, South American, and other countries jointly involved with NASA) landscape and waterscape since the launch of ERTS-1/Landsat-1 in 1972. This impressive archive of environmental color and temperature, thus far analyzed very sporadically and extremely locally for short time-periods, has not been readily available for the time-series of environmental change study originally intended by the NASA visionaries. This is due to the excessive cost of privatized satellite data sale and also the high cost in dollars, manpower, and time of establishing high-speed computer facilities to collect, store, and analyze copious volumes of satellite data.

Nonetheless, as evidenced from Figure 5.15, the predominantly north–south trajectories of Landsat (and other vehicles) readily enables the three decades worth of time-series satellite data to assess changes in environmental ecotones (the transition areas between two ecozones, e.g., between forest and wetland, desert and forest, tundra and plain, etc.). Construction of monthly and/or yearly mosaics of the principal ecotones (and the changes there) would define (or at least present cryptic clues to) the changes occurring in the terrestrial, aquatic, and wetland ecozones subsequent to the launch of ERTS-1. Benefits (economic and social) to the public should be easily presented and understood, and include, but not be restricted to:

- Forestry, agriculture, and resource exploration (sustainability, expenditure, profit) from awareness of changing environmental landscape
- Various business, cultural (lifestyles and traditions of urban, suburban, and native peoples), and entertainment activities intimately linked with and dependent upon climatic and environmental conditions
- Early warning of impending environmental problems (floods; drought; forest and range fires; insect infestations; oil spills; toxic algal blooms; depletion, modification, and/or migration of wildlife habitat) and potential socioeconomic impacts
- Emergence of regions of high and low environmental vulnerability to rate of ecotone change, thereby allowing "hot spots" to become targets for more intensive scrutiny
- Indication of need for initiation of new environmental remediation programs and evaluation of successes/failures of previously established environmental remediation programs
- Provision of a currently absent input to environmental policy formulations, viz., direct time-series evidence of the changing face of the national landscape (input that could assist sensible compromises among cost, disruptions to standard of living, standards for industrial control, socioeconomic mores, taxation, and the necessary limitation, mitigation, and adaptation strategies)

Proposals to perform exactly what has just been recommended have been put forth by many research teams (including EC's NWRI) from many countries for many years. Ecozone mapping exercises for large continental territories such as Canada, the U.S., Russia, Australia, South Africa, Brazil, and others quickly become cost prohibitive when over 30 years of historical archival data are to be obtained. European countries, in which considerably smaller volumes of archival data are required, have fared better (especially countries that display a wide spectrum of distinct ecozones). Space agencies of large continental areas (e.g., NASA, CSA, ESA, ISA) have constructed land cover maps for the U.S., Canada, Russia, India, and other countries. However, these have generally been mosaics of very recent satellite data (from the last year or two) that have, unfortunately, served more as a promotional tool for the launch of future satellite packages and "bookstore wall-hanging products for offices and waiting rooms" than as a tool for environmental assessment.

However, a more successful proposal (for countries the spatial extent of Canada and the U.S.) might require a judicious selection of, say, three north–south transects (west, central, east, or whatever is deemed appropriate) across the country. Certainly the cost of archival data would be dramatically reduced. Assuming an average cost of $750 Canadian per satellite scene and assuming a double pixel purchase for each image center along the orbital path, the 30-year Landsat data archive required for yearly (exclusive of mid-winter) ecozone mapping of three Canadian terrain transects would cost about $2.5M Canadian. Ecozone change information on three selected sectional cuts across the ecozones of an expansive country appears intuitively valuable to public good and environmental policy-making; however, precisely how valuable remains conjecture.

Figure 5.16 illustrates a "net primary productivity" (in units of kgC m^{-2}y^{-1}) land cover map of Canada inferred from AVHRR and Landsat-7 data during 1994. The image was constructed and made available to Environment Canada, other government agencies, and environmental end-users by the Canada Center for Remote Sensing (CCRS) of Natural Resources Canada (NRCan). It is shown here for four reasons:

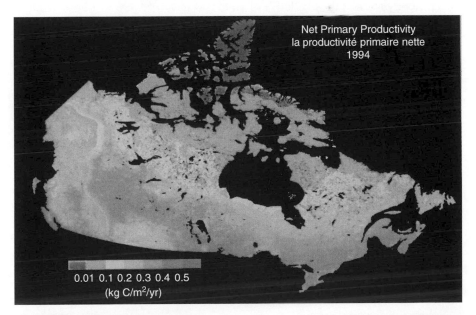

Figure 5.16 (See color insert following page 134.) Net primary productivity (kgC m^{-2} y^{-1}) land cover map of Canada inferred from AVHRR and Landsat-7 data during 1994. This is the terrestrial equivalent to the global ocean productivity map of Figure 5.2. (Public domain courtesy of the Canada Center for Remote Sensing, CCRS, of Natural Resources Canada, NRCan).

- Although certainly qualifying as a picturesque office wall hanging, Figure 5.16 goes far beyond entertainment value and serves to illustrate and promote an important scientific application of environmental monitoring from space: the intercomparison of terrestrial bioproductivity across vast and diverse regions. This figure is the terrestrial counterpart of the global ocean bioproductivity map of Figure 5.2.
- Despite its not being precisely aligned with the ecozones map of Figure 5.15, it does illustrate that primary productivity mapping provides information on the ecotone nature of an environmentally diverse country.
- It illustrates a number of potential terrain transect opportunities for time-series ecozone–ecotone information as discussed in the preceding paragraph.
- It also illustrates (shamefully) that inland waters are invariably represented as "drop-out" data in environmental mapping from space. Ocean bioproductivity and terrestrial bioproductivity are being mapped. Inland water bioproductivity is not being mapped. Sadly, this situation is common practice in virtually every terrain assessment involving space data (other than those utilized in aquatic science research and bio–geo-optical modeling). Filling in the blanks with compatible water quality information and acquiring influential end-users of these filled-in blanks is the charge to us and our readers.

5.13   Seagrass meadows: location and spatial distribution of substrate and substrate vegetation

An emerging focus in coastal oceanology is remote determinations of the spatial distribution and rigor of benthic vegetation (shallow water rushes, indigenous seagrass, and invasive sea algae such as *caulerpa taxifolia* that pose problems in the Mediterranean Sea and *codium fragile* that pose problems off the coast of Nova Scotia) on the substrates of coral reefs, continental shelves, and shallow water coastlines. Although technically not truly grass, seagrasses (silky weed, strap weed, eelgrass) are flowering plants with ribbon-like green leaves. Their roots and horizontal stems are generally buried in sand or mud. The ecology of seagrass meadow bioproductivity is a complex problem that involves seagrass taxonomy, substrate suitability, and water type, as well as the life cycle synergies of the phytoplankton and the aquatic and benthic micro- and macroalgae comprising the ecosystem of a seagrass meadow.

Historically, we have come to regard marine ecosystems as immune to perturbations comparable to those imposed by escalating human populations on terrestrial and inland water ecosystems. Marine ecosystems were considered to be consequences of dynamical stabilities induced by natural climatic phenomena. Seagrass fossils, dating back to the Mesozoic Cretaceous Period (~130 million years BP), have experienced threats of extinction that are hard to attribute to human intervention. In the mid 1930s, *Zostera marina* (eelgrass) essentially disappeared from the North Atlantic due to the so-called "wasting disease." A variety of natural causes was proposed ranging from epidemic-causing bacteria, fungal infection, slime mold, periods of intense and/or reduced precipitation, elevated ocean temperature, and immunity system failure, to sun spot activity. The sudden disappearance of eelgrass had a profound impact on the shallow water ecosystems where it once played a role (e.g., substrate changes from sandy beaches covered with protective vegetation to barren rock; death and/or migration of fish populations).

The eelgrass ecosystems eventually recovered. To the best of our knowledge, however, the wasting disease has yet to be understood fully and therefore might recur. Increasingly, seagrasses are used by humans for such purposes as roofing thatch; fuel; housing insulation; packing material; basket weaving; fiber products; binding of soil and clay in dikes and embankments; manure; and supplement for stock feed. Thus, as human population and human use of seagrasses simultaneously escalate, the threat of impacts to benthic ecosystems from natural processes (not yet fully explainable) is now joined by the threat of impacts from anthropogenic activities (also not yet fully explainable). An urgency has thus been ascribed to increasing our knowledge of the ecology of primary production in seagrass meadows, as well as of the properties and behavior of the substrate material and the water column above it.

Such knowledge will continue to be obtained *in situ* by underwater diving missions. However, because large reaches of continental shelves, coral reefs, and shallow water coastlines now must respond to the escalating world population, underwater diving missions must be complemented by benthic (seagrass) monitoring from space. Remotely sensing visible (from above the air–water interface) substrate material and the vegetative canopy that it might sustain requires, at minimum, numerical values of the following parameters:

- The water-leaving radiance (or remote sensing reflectance) spectrum
- The spectral attenuation by the atmospheric column between the water and the remote sensor platform
- The optical cross section spectra of the indigenous aquatic CPAs
- The total attenuation coefficient of the water column above the substrate
- The hydrographic profile (depth contour) of the substrate
- The optical cross section spectrum (surface spectral reflectivity) of the substrate
- The species and optical cross section spectra of the vegetative cover of the substrate

Bukata et al. (1976) described one of the earliest attempts at estimating hydrographic depth contours from satellite data in which digital band 4 MSS data from Landsat-1 were used to delineate the depth profile of the Mary Ward Ledges along the southern coastline of Nottawasaga Bay in southern Georgian Bay. A mathematical model (and an iterative approach to its usage) was developed to incorporate the substrate reflectivity of these ledges; the optical properties of the water column; the vertical irradiance attenuation coefficient, $K(z,\lambda)$, in the blue region of the visible spectrum; and an open lake atmospheric correction. A similar, more sophisticated approach for hyperspectral sensors (Lee et al., 1999) also incorporates an acquired spectral library of substrate vegetation (West et al., 1985; Laegdsgaard, 2001). This is used at CSIRO (Anstee et al., 2000; Dekker et al., 2003) along with the HYDROLIGHT model of Mobley and Sundman (2000a, b) for satellite assessment of seagrass change.

Excellent textbooks and research-based literature have been generated on the taxa, evolution, and importance of seagrass and other benthic life forms (plant and animal organisms) to the energy transfer mechanisms of the marine food chain. Benthic macrophytes (seagrasses, diatoms, and symbiotic algae in corals and sponges) also contribute to benthic and column bioproductivity. Literature of interest includes, among others, Light and Beardall (1998) and the edited anthologies of Phillips and McRoy (1980) and Birkeland (1997).

In particular, coral reefs are currently the focus of the ecological and remote sensing communities. A large portion of the world's coral reefs is found within the coastal waters of developing nations. The financial constraints of these nations' economies dictate that coral reef management becomes a fiscal onus on developed nations and then becomes a driving force for establishing international markets for space-acquired water quality products. Once again, as has sometimes befallen space-acquired products, the siren call of the seductive incestuous merry-go-round (cautioned against in Section 5.1) irresistibly lures the unwary and the unscrupulous into establishment of apparent end-user interest in a space-acquired private sector product of which the public, despite its unwitting support, is unaware.

5.14 A lost opportunity to present an environmental manager with a compromise between science and usage of science

There is undoubtedly, a plethora of instances that would illustrate the difficulties scientists have in not only communicating with the targeted end-users of their space-derived environmental information products but also with their reluctance to appear "nonscientific" to the eyes of their peers. To illustrate, however, the author needs to look no further than into his own research activities. Thus, for this section, the somewhat presumptuous first-person plural "we" used throughout this book must now demean to an embarrassing first-person singular "I."

As part of its mandate, Environment Canada is concerned with the impacts of fluctuating water levels within the Great Lakes system. Towards this end, Bukata et al. (1988c) presented a conceptual and mathematical model, based upon the geometry of the confining basin, that would predict changes in areal extent of coastal wetlands with persistent changes in ambient water level. Although limited historical Landsat TM data were acquired over northeastern delta region of Lake St. Clair during high water and low water conditions, only the mathematical equations relating coastal geometry to water level were published. Equations tend to neither impress nor interest environmental stewards. However, what was not published might have impressed them (in fact, in private discussions with marsh rehabilitation workers many years later, it certainly appeared to do so). A very simplistic analysis had been performed on three TM images of the Lake St. Clair wetlands (June 20, 1974, and June 26, 1983 were at high water marks; July 14, 1978 was at a low water mark). An essentially trivialized nonsupervised classification was performed on each scene. Only two classifications "water" and "not water" were demarcated. The fear of appearing subsophomoric to your peers is a great deterrent to presenting such blatantly "nonscientific" efforts to an aquatic remote sensing community that was struggling to shed its negative image. Nonetheless, Figure 5.17, offered from the unpublished work of Bukata, Bruton, and Jerome (circa 1985), illustrates the three simplistic "water/not water" nonsupervised classifications of the St. Clare River delta in northern Lake St. Clare. Equally simplistic overlays of these classifications for any two water levels will readily yield a "zeroeth" order map of the areal extent of impacted wetlands (bottom of Figure 5.17). Adherence to scientific rigor is most certainly betrayed by a quasi-subjective demarcation technique that reduces a complex ecosystem to two such exceedingly simplistic segments. However, *relative* change in wetland area is an important factor in wetland rehabilitation efforts. It may, at times, be as important as *absolute* change. We are certainly not advocating that "junk science" be submitted to a serious scientific journal. Nor are we suggesting that overlays of impacted and nonimpacted wetland areas constitute junk science. (*Interjection*: As considered herein, junk science, unfortunately a blatant weapon in the global *warming* controversy, is unfounded opinion by which a sincere albeit gullible, audience is intentionally taken in by spin in place of substance. The overlays of Figure 5.17 do not represent spin. Rather, they represent a first-cut estimate of areal extents of wetlands impacted by persistent water level change. Added scientific sophistication will increase the accuracy of the estimate.) Hence, simplistic activities that could be dismissed as "grade-school science" by a leading scientific journal need not necessarily be summarily dismissed as inappropriate for a specific environmental application. Figure 5.17 may indeed represent an example of the advocated compromise that might open avenues of fruitful discussions amongst providers and users of remotely-acquired environmental information. Thus, due to reluctance on the part of its generators to appear less-than-professional to their peers, Figure 5.17 is an illustration of a possibly lost opportunity to

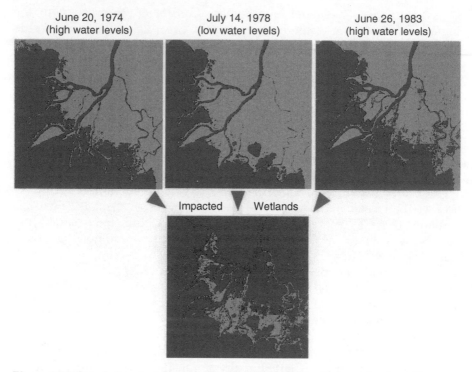

Figure 5.17 Simple Landsat TM classifications of Lake St. Clair wetlands to illustrate the impact of persistent water level changes on the areal extent of wetlands.

promote space monitoring to a dedicated segment of environmental managers and policy-makers.

> I can see no other escape from this dilemma (lest our true aim be lost forever) than that some of us should venture to embark on a synthesis of facts and theories, although with second-hand and incomplete knowledge of some of them — and at the risk of making fools of ourselves.

From *What Is Life*? Erwin Schrödinger, 1943

5.15 Some comments on water quality monitoring

Strains in national, provincial, state, municipal, local, and international relations are stark realities that reinforce the need to have scientifically defendable space products backed by public awareness and acceptance. The onus of making these products the economic motor force falls on the generators of these products. To do so requires that a targeted market be the participants in and power brokers of these "strained" relationships (i.e., federal, provincial, state, municipal, local, and international political policy-makers). The approach that we have suggested here as logical would be from the bottom up, as opposed to the top-down approach currently in place and fraught with problems such as those discussed in Section 5.1 and elsewhere.

The bottom political level of all democratic countries is the general public. The general public (apart from single-issue, self-serving groups that capitalize upon and often monopolize inherent freedoms of democracies) is the most important of all the political levels. However, although it is financing the important and costly activity of space monitoring of environmental resources, the general public is the least influential and least informed player in strained interpolitical relationships. The environmental priorities and issues and the space-derived products discussed in this and other chapters are sound, essential, and beneficial to human and ecosystem health and security. Thus, the value-added public good associated with space-derived water quality products (as well as space-derived terrestrial products) must become known and deemed worthy by the public.

Space monitoring of inland and coastal water quality, on its own, cannot elevate the influence of the general public in alleviating the strain in local to national to international disputes. What it can do, however, is help ameliorate the general public's stature as the least informed player. In Chapter Six we will discuss encouraging signs that the future of remote sensing of inland water resources may be brighter than its past would have led us to anticipate. One of these signs (arising out of justifiable concern for quality of the world's limited supply of freshwater) is a national and international focus on water quality monitoring. To date, however, the water quality monitoring focus remains almost exclusively on operational networks of ground-based stations. As an addendum to discussions in Section 1.7 and as a prelude to discussions in Section 6.2, we close this chapter with another charge that strong efforts be made to merge the protocols and products of space monitoring of inland water quality with those of ground-based inland water quality monitoring networks. Rationale for and consequences of an integrated conventional and space water quality monitoring capability would include, but not be restricted to:

- Water quality, water quantity, and ecosystem health (and the concomitant natural and anthropogenic threats to biodiversity, wildlife, and primary productivity) are inter-related. These essential elements must be integrated, understood, and assessed on a number of spatial and temporal scales.
- Although often fragmented and not intercomparable, coordinated networks of Earth-based water quality monitoring networks exist in several countries.
- Capability to monitor inland water quality from aircraft and satellite sensors exists, but is very highly underutilized.
- Water quality and water quantity monitoring supports a variety of "state-of-the-environment" international, national, provincial, state, and local reporting responsibilities that involve spatial and temporal

trend analyses and local to global research related to climate, land use, and source water protection and security.

- Public outreach capabilities are being gradually developed and implemented through local, national, and international bilateral agreements.
- Value-added monitoring products for political decision- and policy-makers might emerge from successful integrated ground- and space-based monitoring partnerships. Public awareness of the benefits emerging from environmental monitoring might stimulate a national marketplace for space-acquired water quality products. This, in turn, might stimulate an international marketplace for space-acquired water quality products.
- Inappropriate operational protocols, gaps in essential archival environmental data, needs for additional scientific research and/or technology, and possibly questionable locations and data-gathering infrastructure of existing stations along the network can be indicated, addressed, and alleviated.

Throughout this book we have asserted that space monitoring regards itself as neither competition to nor replacement for existing conventional ground-based environmental monitoring. For regions of a large country or waters inaccessible to Earth-based monitoring stations, space-based monitoring may provide the only means of acquiring information on important issues (the Arctic, the Antarctic, oil spills, and dynamic ocean–atmosphere interactive phenomena such as tropical storms, hurricanes, and tsunamis). To obtain directly measured (as opposed to inferred) values of water quality parameters, Earth-based stations are required. Nevertheless, for scientific applications (model development, calibration, and validation) and environmental management applications (assessment, strategies, and policies), EO data from ground-level and above ground-level sensors are interactively valuable.

However, very little monitoring use is made of EO inland water quality products. Very good, although site-specific, work has been the unfortunate norm for such monitoring activities. Ground-based water quality monitoring network operators do not appear to be actively seeking involvement from producers of space-acquired inland water quality products; thus, it would seem obvious that producers of space-acquired inland water quality products must promote their own wares.

> The law of supply and demand has an implicit subclause: that it involves the same people in both capacities. When this subclause is forgotten, ignored or evaded — you get the economics situation of today.
>
> **From *Philosophy: Who Needs It?* Ayn Rand, 1982**

chapter six

Crystal-ball gazing at remote sensing of inland and coastal waters from space

We are all interested in the future for that is where you and I are going to spend the rest of our lives. And remember, my friends, future events such as these will affect you in the future.

From the Edward D. Wood, Jr., film, *Plan 9 from Outer Space*, dialogue by Jerome King Criswell, 1959

Look! Up in the sky! It's a bird! It's a plane! It's...

From the TV series *Superman*, based on the 1938 DC comic-book hero created by Jerry Siegal and Joe Shuster, 1952–1957

Nostalgia implies a longing for long ago and far away times of former happy circumstances. Retrospection, or more appropriately, selective retrospection, is the motor force behind nostalgia. The retrospection that we have been practicing to this point in our discourses, due in no small measure to our concurrent introspection, certainly qualifies as selective retrospection. However, it does not and should not qualify as a motor force for nostalgia. Nevertheless, certain activities and philosophies (epiphanies?), particularly within the recent past, have unfolded that perhaps give hope that some future nostalgia might emerge from some future selective retrospection of space monitoring of inland and coastal water quality. Therefore, in closing this retrospective and

introspective view of space monitoring, our gaze shifts now from
a navel to a crystal ball, where we don our Criswell garb and peer
nostalgically into the future.

6.1   The United States, Canada, and space

"Space" is both a problem and a solution that the U.S. and Canada share.
The large spatial extent and concomitant diversity of North American terrain
prohibit the direct monitoring of even basic freshwater, coastal, and wetland
features from becoming a simple task. Both countries comprise vast territories of widely varying population densities. Canada, in particular, is a vast,
sparsely populated space, with the bulk of its population distributed along
the Canada–U.S. border. Extensive numbers of Canadian lakes, rivers, and
wetlands, as well as continental shelves of three oceans, are far removed
from these highly populated areas and are very expensive to access and
monitor. Water quality and water quantity have been known to fluctuate
quickly and dramatically in both countries, resulting in floods, droughts,
and threats to public drinking water supplies. The river and lake systems of
North America are interconnected through a myriad of inter-relationships.
Thus, the implementation of a policy to monitor this extensive "space"
through Earth observations from "space" would certainly appear to be a
logical and practical solution to this national and international dilemma.

NASA and NOAA deserve full marks for initiating and rendering operational Earth observations (EOs) from space. Extant political driving forces
notwithstanding, the U.S.–USSR space race produced world-class American
space technology and science that could be redirected towards environmental assessment by remote sensing from aircraft and satellite altitudes. Aquatic
optics research performed at American universities and private research
institutes has resulted in adoption of water quality algorithms implemented
in the generation of levels 2 and 3 water quality products derived from
satellite data. Talented workers at the U.S. Naval Research Laboratories in
California, Washington D.C., and the Stennis Space Center in Mississippi
have advanced water quality algorithm development as well as assuming
the Herculean task of alleviating atmospheric intervention issues that are
currently the greatest impediment to EO monitoring of inland and coastal
water quality.

NOAA's CoastWatch Program makes EO data products (also *in situ* data)
available to federal, state, and local water resource managers, scientists,
educators, and the general public. NOAA's CoastWatch comprises several
regional nodes that coordinate processing, delivery, quality control, and
storage of data products from a variety of sensors and satellites in near real
time. The Great Lakes CoastWatch node operating out of the Great Lakes
Environmental Research Laboratory (GLERL) in Ann Arbor, Michigan, is
particularly active in its own operational activities, as well as in scientific

research collaborations with other government agencies, universities, and the private sector within and beyond American borders. Internet-accessible CoastWatch data products include sea surface temperatures, lake and ocean color, and surface winds — products that have contributed directly to such environmental projects as forecasting harmful algal bloom development and dispersal and assessing environmental impacts of near real-time hypoxia (abnormal reductions in aquatic oxygen), species of aquatic nuisance (e.g., sea nettles, exotic mussels), and large-scale climate regulators (e.g., El Niño conditions in the equatorial Pacific Ocean).

Collaborative efforts among aquatic optics experts, academics, the private sector, and public volunteers have evolved slowly, but somewhat surely, over the past several years as attempts to outreach the knowledge accumulated (and accumulating) from the scientific community to its intended end-users. Included is the training of students and professionals. One such collaborative entity is the Student Mentored Advanced Research and Technology Satellite (SMARTSAT) program initiated by Stanford and Washington/St. Louis Universities, NASA, Skywatch Information Systems, and other partners; the mission is to provide training and employment for students in obtaining and interpreting space data. A second such collaborative entity is NASA's Regional Earth Science Applications Center (RESAC) and Affiliated Research Center (ARC), consortiums of academic and federal agencies and industries in Minnesota, Wisconsin, and Michigan; the mission includes developing EO water quality products and outreaching their end-user value by directly engaging the public in the product development.

Canada, which became the third nation in space with its 1962 launch of the telecommunications satellite Allouette, deserves some good marks too. Since its maiden voyage aboard the U.S. space shuttle *Columbia*, in 1981, the Canadian-developed remote robotic manipulator system known as Canadarm has been an invaluable "grabbing arm" for satellite orbital placement and satellite system repair within the American space shuttle program. Canada was the first country to form an EO partnership with the U.S. to track the NASA ERTS-1/Landsat-1 space vehicle independently over Canadian terrain; it also provided coinvestigators for the American Skylab Program.

Throughout the 1970s and 1980s, the Canada Center for Remote Sensing (CCRS) of Natural Resources Canada (NRCan) was the major force behind the promotion of EO in Canada. At its outset, CCRS adopted a user-friendly policy with the scientific community self-charged with obtaining and analyzing satellite data. Under this policy, end-users could acquire the Landsat data for a modest cost-recovery fee. (This policy allowed NWRI to pioneer the use of digital ERTS computer-compatible magnetic tape data to produce very quickly analyses such as the bifurcated sediment transport model origin for the Point Pelee Rondeau landforms described in Section 5.11.) However, following NASA's lead, the CCRS experiment of commercialization of Landsat data sale became an irrevocable failure. The ensuing rapidly soaring costs of Landsat data and their financial inaccessibility to the scientific community dramatically inhibited the development of applications of EO. This long-standing policy of data inaccessibility has been responsible for the three decades of archived environmental data that lie unanalyzed to this day.

Recently, the U.S. government has resumed control of the Landsat data and, as a result, the cost of Landsat-7 data has dropped significantly. Also, again following NASA's lead, CCRS adopted the handicapping policy of "all-inclusiveness" usership (along with a multitude of pilot projects apparently designed to illustrate the value of remote sensing to a series of environmental problems that could be solved without remote sensing); this added to the burdensome cross of suspect reputation already borne by the remote sensing community. An unintentional ironic assault on this all inclusiveness was invoked by the CCRS decision to place all of Canada's remote sensing "eggs" into a single "basket" of all-weather microwave radiation.

Radarsat-1, Canada's first EO satellite, was launched on November 4, 1995. It carried an active synthetic aperture radar (SAR) sensor sensitive to the C-band wavelength of 5.6 cm. This focus on a SAR space vehicle was designed to capture Canadian prestige within the international space community. The focus was applauded and endorsed by Environment Canada's Atmospheric Environment Service (now Meteorological Service of Canada, MSC). MSC is the major Environment Canada user of space-acquired data. Launch of Radarsat-2 is imminent. Sadly, the CCRS commercialization policy has continued with Radarsat data, along with, not unexpectedly, significant resentment. MSC notwithstanding however, across Canada as a whole, only a few individuals in government, university, and the private sector are using the radar data.

In fairness, Radarsat has proven of value, although often limited value, to the weather-related interests of end-users and to water-cycle and hydrology-minded scientists concerned with water quantity (floods, droughts, glaciers, tundra, snow-water equivalents, ice pack, and precipitation). Radar frequencies have also been used in the surveillance of Canada's 200-nautical-mile exclusive economic zone (EEZ) and form the basis for Canada's Integrated Satellite Targeting of Polluters (I-STOP) Program. However, the CCRS focus on a limited end-usership by the CCRS sole focus on radar and, thus, exclusion of the rest of the electromagnetic spectrum was understandably regarded as betrayal by the bulk of the scientific remote sensing community concerned with EO as a tool for ecosystem monitoring.

With environmental "color" falling off the CCRS "radar screen," inland and coastal water quality scientists became essentially disenfranchised from Canadian EO efforts. Thus, in a variety of ironic ways, the "all-inclusiveness" problem of public acceptance of EO from space became a problem of "all-exclusiveness," with aquatic optical elitism and inland and coastal water quality monitoring being principal casualties of this newly-adopted remote sensing directive.

In 1989, however, the Canadian Space Agency (CSA), an organization that falls under the auspices of the Canadian Ministry of Industry, was established. Also, Canada became an associate member of the European

Space Agency (ESA). The CSA quickly realized that, in order to develop and use space technology to acquire national and international markets for the products emergent from Earth observations from space, satellite data collected at other than SAR frequencies were also necessary and that environmental assessment was a key to the end-usership of EO data products.

The CSA was also cognizant that, apart from the large-scale EO activities of Environment Canada's MSC meteorological (atmospheric, weather, and climate) operational activities (heavily weighted towards Radarsat), EO activities throughout the rest of Environment Canada (water, wildlife, and wetland) were minuscule and fragmented. Very small numbers of highly specialized experts (with nonexistent or at best very limited space-data manipulation hardware and software facilities) were thinly scattered across Canada. Therefore, these researchers were essentially forced to become expert theoretical modelers; as such, they depended upon the private sector to generate EO products to extend or validate their own models.

As discussed in Section 1.5, the scientific community is its own end-usership. Clearly, this is not conducive to mobilizing a dedicated end-usership for EO applications. However, the CSA's mandate revolved around that of the Ministry of Industry. Therefore, development of partnerships among the business community, the EO science community, the environmental monitoring community, and the CSA quickly loomed as a consummation devoutly to be wished. Steps taken to achieve such consummation are discussed in Section 6.2.

Consistent with the marketplace reality check discourses of Section 5.1, the strong and growing feeling within developed nations is that satellite data should be freely available because costs of sensor development, satellite launching, data acquisition, and raw data processing have already been paid for through public taxation and that governments should bear the operational costs of products for public good. An obvious desirable by-product of such time-series data availability would be the stimulation of time-series environmental research and development of environmental indicators and stress–response functions.

As an illustration, the use of AVHRR (despite, or maybe because of, its simplicity) is very well utilized in a variety of agricultural, forestry, and water applications largely because of end-user access to a wealth of continuous and free sources of data. The AVHRR satellite series was designed to acquire weather data. International agreements on weather data exchange have been instrumental in this free access to data and have also allowed research-minded institutes to establish dedicated tracking stations to acquire AVHRR and SeaWiFS data. As data from satellites of similar data capture situations become competitive with private sector monopoly, there should be some optimism that pricing policies will slowly divest of the unrequited greed that seems to guide them. To date, however, data vendors continue to dismiss the "cost per data bit" as opposed to the "cost per image" pricing policy suggested by Bukata and Jerome (1998).

Prior to closing this section, it would be irresponsible not to recognize the important contributions of two organizational entities. Although the project offices of each are housed within Canada, their membership and outreach impacts are international.

The first is the International Ocean-Color Coordinating Group (IOCCG) principally housed within the Bedford Institute of Oceanography (BIO) at Dartmouth, Nova Scotia. Established in 1996, the IOCCG is, in part, financially supported by national space agencies. Through collaborative efforts among internationally recognized aquatic science experts, it advocates and promotes practical applications of remotely sensed mid- and coastal-ocean color data by coordinating fragmented water-color programs; by providing training and expert advice; and by liaising between providers and users of space-derived water quality and water quantity products. Its insightful reports (particularly IOCCG, 2000, which deals with the science and applications of remotely monitoring optically complex waters) are invaluable. Its most recent report, IOCCG (2004), focuses on approaches to interutilize, in a long-term archival manner, ocean color data collected from a galaxy of sensors of nonidentical spectral and spatial resolution. A variety of techniques is used for differentiating, averaging, and accumulating individual pixel values of diverse data sets into appropriate "bins." These "binned" pixel values are then used to generate level 2 and level 3 products from satellite-specific, nonidentical models and algorithms.

Sidebar

Numerous differences in remotely acquired aquatic data sets make their direct intercomparison problematical (disparate spectral and spatial sensitivities of space sensors, their viewing geometries and calibrations; specific corrections to improve data integrity; generation of levels 2 and 3 water quality products from various algorithms; nonidentical mathematical criteria for acceptance or rejection of specific pixel values; techniques for averaging or normalizing accepted pixel values; and storing outputs of such techniques). The general approach to coping with these and other disparities is to compile, where possible, data into "bins" that best define pixels whose individual numerical values can be intercompared in a manner that maintains the integrity of the comparison.

A principal goal of launching specialized suites of space sensors is to build a long-term, internally consistent data base of environmental (terrestrial as well as aquatic) information from which short-term and interannual change may be interpreted (within a well-defined geographic grid system) in terms of climatic forcing functions. Binning, therefore, is an activity that requires high reliability if environmental monitoring from space is to analyze spatial and/or temporal environmental change effectively. The mean annual global field of primary production (Longhurst et al., 1995) illustrated in Figure 5.2 is an illustration of skillful use of reliable binning.

Binning algorithms for SeaWiFS may be found in Campbell et al. (1995). Binning schemes are also available for production of level 3 products from OCTS, CZCS, MODIS-Terra, MODIS-Aqua, ADEOS, GLI, and other water-viewing satellites. Although binning is important for space monitoring of ocean, inland, and coastal water quality, it is equally important for terrestrial and wetland monitoring from space. Ground-based terrestrial and aquatic monitoring networks have comparable data-binning requirements.

The second organizational entity is the Alliance for Marine Remote Sensing (AMRS) located in Halifax, Nova Scotia, which is also international in its membership. Operating under a currently restrictive corporate financial sponsorship, the AMRS has nonetheless established important thematic workshops and produced invaluable literature and water-product advocacy through its official magazine, *Backscatter*. This publication showcases applications of work by internationally recognized scientific and technical authorities. The IOCCG and the AMRS were initially conceived as purely oceanographic in their foci (as their names declare). Although each retains its marine focus, the recent spotlight on coastal zones has incorporated case 2 waters into those foci. The AMRS, in particular, has focused quite intently on promoting the value of space-acquired products of inland water quality to influential end-users. The 82-page winter, 2004, special issue of *Backscatter*, which is thematically devoted to Canadian Space Agency Program activities, is a valuable read.

6.2 Addressing the chasm between Canadian inland and coastal water quality products and their potential end-users

It is trite to restate that major opportunities for Canadian research and operational activities in ground-based and space-based EO of inland and coastal water quality (see Section 4.4), as well as for marketing of industrial water quality products emergent from those activities, exist or should exist. Canada has the longest coastline in the world, over 10% of the world's freshwater supply, over 20% of the world's standing lake water, and nearly one-quarter of the world's wetlands; thus, one would assume Canada's need for monitoring, mapping, and managing these natural resources to be intuitively obvious. Monitoring the quality of standing lake water would appear all the more essential when it is realized that the totality of global standing lake water represents only ~0.14% of the world's water supply. Thus, Canada's ~0.03% of the world's water supply further underscores the need to monitor and protect this limited, essential, and precious resource.

Over the years, Environment Canada (EC) has established a ground-based network of operational water quality monitoring stations. This network is highly fragmented and dispersive, thereby offering somewhat limited syntheses of integrated Canada-wide, basin-wide, or regional water quality. Nonetheless, it appears to be moving in a positive direction in collection and synthesis of water quality information as well as in its subsequent dissemination to environmental stewards and the general public. It also appears motivated to provide some form of "water quality indicators" upon which reliable appraisals of use and abuse of Canadian waters can be continually made. Although such indicators are still essentially conceptual, the intention to give concept form is encouraging.

Conspicuously absent from EC's cross-country water quality monitoring protocol, however, are space observations of inland water quality. Harking to a recurrent theme of this book, it should be obvious that inclusion of space observations (not just in Canada, but across the entire globe) is eminently sensible. Perhaps, however, an alternate approach to defense of this recurrent theme might be considered. Protection and conservation of environmental resources and aquatic ecosystem integrity require recognition of the myriad inter-relationships of inland water quality with human health and ecosystem health (e.g., comfort, recreation, security, economy, biodiversity, wildlife, threats to water supply, etc.). A logical coordination of ground-based and space-based inland water quality monitoring provides an opportunity to obtain direct information on the sustainable interactive dynamic resulting from human and wildlife populations coexisting within the environmental framework of a common freshwater basin.

Water quality is essential to the health of human and indigenous wildlife inhabitants. Ground-based and space-based EOs of water quality adroitly dovetail because of their similarity and dissimilarity. Due to their similarity, space determinations of aquatic and terrestrial change enable the point source data acquired by the ground-based network to be extrapolated to and displayed upon a larger spatial scale. However, dissimilarity in water quality parameter monitoring capabilities between direct and remote water quality monitoring might also be advantageous. Concentrations of toxic biological and chemical contaminants can be readily determined by laboratory analyses of ground-based lake, river, and groundwater samples. Water quality threats to human health, therefore, fall logically within the domain of the ground-based networks.

Remote sensing cannot monitor an aquatic toxicant concentration, even if the severity of the concentration is fatal to human health. However, remote sensing can monitor aquatic and terrestrial chlorophyll concentrations and, therefore, *relative* if not *absolute* changes in aquatic and terrestrial bioproductivity on scales the size of large watersheds. Thus, combining ground- and space-based water quality data with point and raster data of terrestrial land use should provide insight into the interactive issues of human health, eco-system biodiversity, and environmental status of the watershed.

Aside

Combining space-based lake raster data with ground-based basin point data suggests a possible way of making remote sensing more "palatable" to water managers. The suggestion revolves around two totally different (although certainly relatable) approaches to applying space data to water quality issues:

- The way in which we have normally been tackling the problem — namely, using forward and inverse modeling to generate "maps" of near-surface chlorophyll, inorganic turbidity, clarity, primary production, etc. to illustrate some aspect of temporal and/or spatial changes in near-surface water quality
- Remotely monitoring terrestrial aspects of the confining basin or watershed and using temporal changes therein to account for changes in water quality that have been directly observed within ground-acquired data

Cause–effect relationships of inland water quality and land use are generally predicated (although there are exceptions) upon land use being the cause and inland water quality being the effect. Therefore, if the combination of space-based lake raster data and ground-based land-use point data is a valuable commodity, then combination of space-based lake raster data, ground-based lake point data, space-based land-use raster data, and ground-based land-use point data becomes even more valuable. An associated corollary to the second approach to utilization of space-acquired watershed data logically revolves around the question, "Can remotely monitoring nonwater features be of equal or even more consequence to case 2 water quality monitoring than remotely monitoring water features?" Intuitively, the answer would appear to be "probably not." However, under certain conditions (atmospheric, logistic, geographic, historic, absence of relatable data-sets, etc.), such analyses might be the only interpretative avenue available. Certainly, all the methods briefly discussed here should be considered in advancing watershed monitoring from space as a practice that water quality managers and policy-makers would find effective in addressing their mandates.

Issues of the inter-relationships of watershed land use and lake water quality are currently addressed through collaborative government and private sector projects under the aegis of the CSA's Earth Observation Applications Development Program (EOADP) and Government Related Initiatives Program (GRIP).

EC and Fisheries and Oceans Canada (DFO) are research-oriented federal government agencies mandated to conserve and protect environmental resources and aquatic ecosystem integrity. They share a common interest in monitoring the quality (i.e., color) of optically complex water bodies. Space-based observations of water color provide a major (and often the only) window into freshwater and marine biology on synoptic scales. DFO is actively utilizing both Radarsat, in order to apply EO to its water quantity and ice-dynamics mandate, and full-spectrum water color space data, to apply EO to its mid-ocean and coastal ocean water quality mandate. However, along with CCRS, EC is almost exclusively utilizing Radarsat data to apply EO to the Meteorological Service of Canada (MSC) mandate of atmospheric physics, weather forecasting, hydrology, and snow–ice dynamics. Nonradar frequencies are used by EC's National Water Research Institute (NWRI) in Burlington, Ontario, for quality of Great Lakes and other inland water bodies, and EC's Canadian Wildlife Service (CWS) for wetland and wildlife habitat mapping.

In an attempt to surmount the barriers inhibiting the exploitation of space-acquired water quality products for environmental management and policy formulation, NWRI (of EC), DFO, and the CSA have responded by novel model validation demonstrations, dialogue among providers and end-users of space-derived water quality products, and establishment of integrated remote sensing programs involving government, university, and private sector colleagues. A few of the turning points in the odyssey of NWRI, DFO, and the CSA in addressing the conjoint government, academe, and industry obstacles that preserve the chasm between space-acquired water quality products and their intended end-usership included (but was not restricted to):

- In the mid-1990s, inland water quality monitoring was incorporated into the CSA's Long-Term Space Plan (LTSP), providing recognition of freshwater systems as an integral aspect of the Canadian landscape as seen from space.
- The 1997 Canada Oceans Act, along with the earlier Federal Water Policy and recent Canadian Innovation Strategy, collectively entrenched aquatic ecosystem approaches to ocean and freshwater management and thus provided government justification for space determinations of primary production of natural waters.
- The advent of the era of the hyperspectral remote sensing (ushered in by the launch of the Earth Observation Platform, EOP-1 with its sensor Hyperion) and the CSA's involvement with ARIES, ENVISAT/MERIS, SmallSat, and others provided access to the full-color

spectra essential to water quality model and multivariate optimization methodology.

- CSA has planned development and launch of a Hyperspectral Environmental Resource Observation (HERO) sensor applicable to forestry, agriculture, geology, coastal and inland waters, and environmental monitoring in general
- CSA has developed (along with government, university, and private sector scientists) a cross-country coordinated Canadian Expert Support Laboratory (CESL) to asses the capabilities of the European Space Agency's satellite sensor, the Medium Resolution Imaging Spectrometer (MERIS).
- CSA has established a Canadian Hyperspectral Users and Science Advisory Team to develop environmental products that would provide national and international markets for Canadian business.
- CSA has awarded contracts to the Canadian private sector to provide, by direct contact with environmental stewards and the EO scientific community, water quality management priorities; data-collection practices; availability of information products and services; roles of existing and planned EO satellite missions in meeting information needs; and identification of barriers to the acceptance of water quality products from satellite data.
- Dedicated satellite tracking stations have been established for the capture and analyses of POES/AVHRR and SeaStar/SeaWiFS optical data at DFO's Institute of Ocean Sciences (IOS) and EC's NWRI, in addition to nongovernment Canadian institutes focused on aquatic research.
- The International Ocean-Color Coordinating Group's comprehensive Report No. 3 (IOCCG, 2000) summarized applications of state-of-the-art remote sensing of inland and coastal waters and was directed to scientists, students, water managers, policy-makers, and space agencies.
- In 2001, the Remote Sensing and Resource Management in Near-Shore and Inland Waters Workshop in Wolfville, Nova Scotia, sponsored by the Alliance for Marine Remote Sensing (AMRS), the CSA, National Resources Canada, and others, in which a forum was established to address end-user disinterest in space-derived EO water quality products. Invited CSA managers; government, university, and industry scientists; believers; nonbelievers; skeptics; vendors; government managers; and users and potential users of water quality products attended the forum. A similar forum was invoked at the concurrent scientific conferences of the AMRS and TOS (The Oceanographic Society) in New Orleans, Louisiana, in 2003.
- The NWRI gave a presentation on the applications of EOs of inland water quality to the Canadian government's environmental priorities to the Nature Table of EC.

- With direct input from EC and DFO, CSA established the Earth Observation Applications Development Program (EOADP). Of particular value to water quality monitoring were CSA contracts let to (1) G.A. Borstad Associates Ltd., the NWRI of EC, and the IOS of DFO; and (2) the Freshwater Institute (FI) and IOS of DFO. Both contracts were to develop and promote useful water quality products for inland and coastal Canada.
- CSA established the Government Related Initiatives Program (GRIP), including, in particular, the projects "A Wealth of Water: Wise Management of Our Watersheds, Lakes, and Rivers Systems" (water quality monitoring of inland waters) with EC/NWRI and "The Ocean's Pulse: Moving towards Operational Use of Ocean Color" (water quality monitoring of coastal ocean) with DFO.
- Special issues of the Alliance for Marine Remote Sensing (AMRS) official magazine, _Backscatter_, have been consistently devoted to the science and the promotion of inland, coastal, and mid-ocean water quality, including periodic editorials directly aimed at the consistent lack of end-usership interest (e.g., Bukata, 2001; Whitehouse, 2002).
- Senior management of EC has shown a growing desire to define and thus develop a cross-country set of water quality indicators that would most assuredly need to involve EO of inland and coastal water quality.

The EOADP and GRIP water quality projects are discussed in Section 6.3.

Thus, although a chasm-spanning bridge (linking space-acquired water quality products to a sought after end-usership) is being constructed in an ofttimes sporadic and piecemeal manner, the chasm remains. To span the chasm fully, however, requires more than compelling illustrations of the value-added nature of space-derived water quality products by the scientific community that has generated them. It also requires a political will to overcome the deeply rooted mind-sets that might have relegated environmental monitoring from space to a "holding cell" with possibly no "faint hope clause" for parole.

The key gaps in operational water quality mapping from space are the availability of and access to satellite color data and a tenuous market for water quality products. Historically, environmental decision-makers and political policy-makers within EC have shown general disinterest in remote sensing. Apart from NWRI, DFO, CWS, CSA, academe, and the private sector, very little environmental monitoring interest in the Landsat archives, MODIS, MERIS, SeaWiFS, HERO, Hyperion, and hyperspectral sensors in general is shown. The sole focus by the CCRS, Natural Resources Canada, and EC has been and is on Radarsat and SAR. However, cracking this microwave monopoly on the mind-sets of end-users and finding chinks in the armor of the historic disinterest now appears a little less improbable with the CSA as an ally than it did a decade ago with the CCRS as a formidable obstacle to color-based sensors.

Many of the activities of the combined efforts of the CSA, NWRI, DFO, and their private sector and academic scientific colleagues have been tried throughout the past three decades, with little or no success. However, although the script may be familiar, new players are in the cast; these players no longer feel that they need resort to the suspension of disbelief that was imposed by earlier proponents of the applications of EOs from space. After a prolonged hiatus, an "elitism" of environmental monitoring from space is re-emerging (in North America and globally), slowly replacing the earlier rampant and restrictive "all inclusiveness" and regaining the influence from which it had abdicated. Such elitism may, in time, produce a respected pedigree for water quality monitoring from space. Once a pedigree is in place, space monitoring, like pedigreed disciplines and activities before it, should be capable of withstanding inappropriate work (i.e., the reputation of the bad workers rather than the reputation of remote sensing, will suffer). Whether the current cast will succeed where the original cast failed, however, remains moot. Nevertheless, to quote an undocumented quip from a forgotten philosopher, "A fine start is a good beginning."

6.3 EOADP and GRIP

Some objectives of the CSA's EOADP and GRIP are

- To develop *useful* space-acquired water quality products for inland and coastal Canada
- To promote, through these products, end-user awareness of the ability of hyperspectral satellite missions to provide useful water quality information
- To establish, for the private sector, national and international markets for these products

A logical tripartite collaborative effort emerged among scientific colleagues from government (NWRI of EC, IOS of DFO) and the private sector (G.A. Borstad Associates Ltd.) for first establishing the inland and coastal water quality directives of the EOADP, then conducting spectro-optical projects under the CSA sponsorship of EOADP and GRIP. Pertinent to both program projects were generation and intercomparison of case 2 water quality products using

- Full-color bio–geo-optical modeling and multivariate optimization analyses
- *In situ* determinations of downwelling and water-leaving radiances and the inherent optical properties of indigenous color-producing agents
- Fluorescence line height (FLH) methodologies on Laurentian Great Lakes waters and the Canadian Pacific Ocean coastal region

Specific to the EOADP water quality program were compilation of satellite image (MODIS) and *in situ* optical data bases (NWRI suite of above- and below-surface spectrometers under the acronym WATERS, for water–air transference of electromagnetic radiation). These were used to intercompare measured and modeled remote sensing reflectance; measured and modeled aquatic component concentrations; multivariate optimization and FLH methodologies; and ground-level measured and satellite-recorded level 1 and level 2 data products. Initial results of the project were presented by Borstad et al. (2002a, b). Lessons learned from the tripartite EOADP project are discussed in Bukata et al. (2003).

Two GRIP programs have been initiated that appear to bode well for recognizing value-added nature of space-derived inland and coastal water quality information. The CSA's "A Wealth of Water" project is designed to generate water quality maps of the lower Great Lakes in Ontario and Lake Winnipeg in Manitoba from MODIS, MERIS, and SeaWiFS data. These EO observations of lake water quality are related, using GIS methodologies, to locally acquired land-use data sets within respective watersheds (including river inputs and groundwater data). In addition to environmental impact assessments, this project is designed to address the dearth of entries in the catalogue of optical cross section spectra of indigenous CPAs and the integration of EO from space into Environment Canada's EO network of ground-based water quality stations.

The CSA's "The Ocean's Pulse" project is designed to build upon DFO's water quality programs in the Atlantic, Pacific, and Arctic Oceans and advance aquatic ecosystem modeling by coupling water color into impacts of intense tropical storms on aquatic bioproductivity. These two inter-related GRIP projects are described by Bukata and Helbig (2004). Hopefully, the EOADP and GRIP visions are building the infrastructure in advance of the environmental monitoring capability of the CSA's proposed hyperspectral satellite HERO. Current EOADP and GRIP experience and emergent water quality products might play significant roles in upcoming national and international EO programs (e.g., EOS of Section 5.1).

Borstad Associates has made a major contribution to the NWRI effort to outreach the value of space-derived inland and coastal water quality products to students and the general public. The group has created an ingenious user-friendly temporal profiler of historical SeaWiFS data to produce, from NASA algorithms, "instant" time-series maps (and movie looping of them) of a variety of optical and water quality variables for large lakes in the northern hemisphere. The profiler is being extended to include southern hemisphere lakes and water quality information garnered from other satellites.

Although the two preceding projects are most pertinent to the contents of this book, existing and planned GRIP projects extend to integration of terrestrial, aquatic, and wetland ecosystem bioproductivity under the distinct, yet interactive, EO directives of (1) environmental monitoring, science, and prediction; (2) mapping, inventory, and resource management; and (3) safety, security, and law enforcement.

6.4 The compact airborne spectrographic imager (CASI)

Of all the airborne imaging sensors that collectively cover a substantial portion of the electromagnetic spectrum, the most universally applied and valuable to water quality monitoring (along with numerous terrestrial applications) would arguably be the Compact Airborne Spectrographic Imager (CASI). As such, its historical and current usage warrants a special citation. In an earlier book (Bukata et al., 1995) we bemoaned the lack of CASI-type sensors aboard existing or scheduled environmental satellites. We need bemoan no longer. Since the late-1980s, the concept of imaging spectrometers (see the excellent treatise by Dekker, 1993) and hyperspectral imaging of environmental color have become increasingly popular. Terrestrial and aquatic CASI data appear to have served successfully as a springboard to current and planned hyperspectral satellite environmental monitoring missions.

The CASI is a push-broom imaging spectrograph comprising a two-dimensional CCD (charge coupled device) array. A fully programmable quasihyperspectral airborne sensor developed in the late 1980s and produced by ITRES Research Limited of Canada, CASI records electromagnetic radiation in the 400- to 1000-nm range. A number of ITRES CASI units have been modified and used by the private, public, and academic sectors in several countries for a variety of research applications. It collects data (in real time) in a choice of spatial (up to 18 spectral bands with a swath of 512 pixels of sizes generally 2 to 4 m) or spectral (generally up to 288 spectral channels but with decreased spatial resolution) operating modes. CASI data can be readily geometrically registered and, by means of specialized image-processing software, provide diagnostic information regarding the absorption and scattering properties of a terrestrial or aquatic target.

Over the years, applications of CASI data have, through theoretical and applied research, included mapping of wetland vegetation; identification of mineral composites; classification of soils; identification and rigor of cultivated and natural vegetation; mapping of near-shore benthic communities and bathymetry profiles; and, recently, mapping the quality of inland and coastal waters. Such synoptic environmental products considered as time series have contributed to inland water and coastal zone management.

The variety of readily selectable operational modes of the CASI instrument (up to a full-frame image employing all 512 pixels and all 288 spectral channels) plus a judicious selection of altitude and local weather-related flight plans enables implementation of features of terrain monitoring protocols that are denied the repetitive terrain coverage of Earth-orbiting satellites. Two such advantageous features are:

- It offers at least partial control of intervening atmospheric conditions by (1) judicious selection of direct solar/diffuse sky radiation conditions for overflights; and (2) on-board, real-time sensors to estimate atmospheric attenuation coefficients recording properties of a smaller atmospheric path and one directly linked to the target. Further, multialtitude flight paths (striated or spiral) may provide added information on the impact of atmospheric aerosols on the color spectrum recorded at the aircraft.
- The flexible programmability into predetermined suites of spectral and spatial configurations has enabled CASI spectra to simulate environmental products that might be anticipated from planned environmental satellites.

This simulation capability has been (and is being) utilized to obtain previews of quantitative digital space mapping of freshwater and marine targets. Perhaps because an ancestor of the current CASI was an airborne fluorescence line imager (FLI), CASI has been used to preview an anticipated MERIS product based on chlorophyll fluorescence. Figure 6.1 (provided courtesy of G.A. Borstad Associates Ltd.) illustrates a simulated MERIS surface chlorophyll map from a CASI overflight of the Bay of Quinte (on Lake Ontario and stretching north to the Trent River Valley) on September 21, 1992.

Algorithms considered for use in generating such chlorophyll concentration maps from CASI data are discussed by Clarke et al. (1970); Neville and Gower (1977); Gower and Borstad (1981); and Gower (1999). The Bay of Quinte area, with its large range of autumnal chlorophyll concentrations

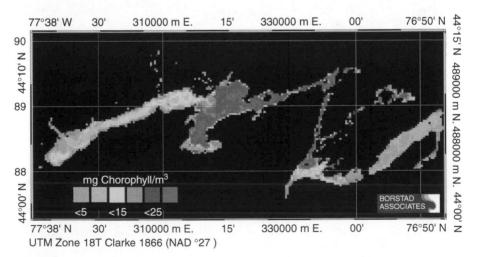

Figure 6.1 (See color insert following page 134.) Simulated MERIS surface chlorophyll map from a CASI overflight of the Bay of Quinte (on Lake Ontario and stretching north to the Trent River Valley) on September 21, 1992. (Provided courtesy of G.A. Borstad Associates.)

(<5 mg m^{-3} to >25 mg m^{-3}), is an excellent choice of aquatic target to simulate anticipated space chlorophyll products not only from fluorescence models, which focus solely on red wavelengths, but also from multivariate optimization models, which consider the full-color spectrum (see discussion in Section 6.7 and Figure 3.1).

Certainly our preference for CASI as a prototype for satellite sensors directed towards recording water color is obvious. However, other airborne imaging spectrometers (e.g., AVIRIS, AISA, HYMAP) have also been used for obtaining and simulating aquatic products and, therefore, deemed appropriate as satellite sensor prototypes. Since the mid-1980s, the concepts of airborne determinations of high-definition spectral return from aquatic resources have benefited studies of inland water quality (organic and inorganic) and coastal ocean planktonic blooms, as well as providing increased knowledge of coastal processes, bathymetry, coral reefs, continental shelves, and inland and coastal wetlands.

6.5 Directions taken at CSIRO

Australia's Inland and Coastal Remote Sensing Group at the Commonwealth Scientific and Industrial Research Organization (CSIRO) has recognized the needs to elucidate the relevance of space-acquired aquatic information to managers charged with environmental stewardship and to harness current technologies to assist elucidating this relevance. This group has been focusing upon environmental issues associated with sensitive coastal ecosystems such as seagrass, mangroves, and other submerged aquatic habitats. CSIRO's earlier work successfully used Landsat MSS and TM data to map the Great Barrier Reef; CSIRO was also one of the first research institutes to "road test" the water quality applicability of NASA's hyperspectral sensor, Hyperion, aboard the EO-1 satellite (Dekker et al., 2002b). Over the years CSIRO has also used CASI data to map aquatic chlorophyll, turbidity, and deleterious algal blooms on a global basis.

Scientifically defendable, application-oriented space observations of natural waters are proceeding with promise at CSIRO — testimony to its enviable history of research excellence. Specifically, the aquatic optics research community owes a large debt of gratitude to several decades of immaculate theoretical modeling by John Kirk (for a capsule look, see Kirk, 1994a, and then his myriad milestone papers). It is comforting to see his legacy being merged with a newly developing legacy of user-oriented space applications. As an illustration of CSIRO's realization that end-users must be made aware of the value of space-derived inland water quality products to their monitoring priorities, Dekker et al. (2001c) developed a generic methodology based upon sound modeling of the water–atmosphere system to provide end-users with adequate water quality information at an appropriate time and format and at a competitive price.

A recently implemented CSIRO program addresses stakeholder requirements for regular updated maps of coastal water quality (chlorophyll and sediment loads) and aquatic vegetation (seagrass and algal blooms). Instructional and readily understandable toolkits that are designed to enable scientific and nonscientific stakeholders to incorporate these mapping techniques into their coastal zone management practices are being prepared.

Coastal ocean waters are, of course, of local and global relevance because they support fisheries, provide natural spawning grounds, and are increasingly utilized in industrial aquaculture. Combinations of cyclical climate-related impacts (e.g., erosion, accretion), human activities (e.g., tourism, recreation), and blatant overfishing have placed coastal waters and subsurface reefs and shoals in jeopardy, as well as seriously depleting food fish stocks (e.g., Atlantic cod, Pacific salmon). Updated mapping of coastal water quality provides much needed information on the presence and temporal evolution of these (and other) impacts. Clever analyses might then provide information on the nutrient–bioproductivity and predator–prey dynamics invoked by these impacts (à la Platt et al., 2003). CSIRO programs encouraging public participation (and therefore addressing in a most positive manner the need to bring the sponsoring public aware and onside) provide reinforcement to the public-involvement policies of the RESAC/ ARC water clarity programs discussed in Section 5.3 and the public policing of fishing reserves violations proposed by Clover (2004) that is discussed in Section 5.6.

6.6   Suspended sediments recognized as independent CPAs?

For considerable time we have been advocating (Bukata et al., 1981a, b, 2004) the use of four or five component bio–geo-optical models for inland waters in which indigenous suspended minerals, which are surrogates of choice for inorganic sediments of terrestrial origin, are considered independent color-producing agents (CPAs) of optically complex waters. It would seem eminently sensible to determine and catalogue the absorption and scattering optical cross section spectra (IOPs) of a variety of representative aquatic sediments because

- In most turbid waters, suspended sediments are the dominant CPAs.
- In turbid waters where they are not the dominant CPAs, suspended sediments are competitive CPAs that modify the color that water would display in their absence.
- These terrestrially derived inorganic particulates are a consequence of geologic diversities of the confining basins.

As discussed in Chapter Two and Chapter Three, such a catalogue has contained very few entries. Largely due to their absence from pelagic marine waters, the historic focus on ocean color, and the relatively late-start of focus on lake color, suspended sediments have become, as we described them in the opening comments of Chapter Three, the "somehow 'forgotten man' in the saga of case 2 water science." However, recently the number of direct determinations of inherent optical properties of indigenous suspended inorganic matter has increased. Perhaps this is due to the relatively recent focus by the ocean optics community on coastal zone waters and its realization that the lake optics community is struggling with a problem from which it had hitherto been immune. Perhaps this is due to the realization that inorganic matter cannot be accommodated within bio-optical models by merely adding a backscatter term. Perhaps this is due to a recent IOCCG workshop in which the dearth of inorganic IOPs was an issue.

Whatever the reason(s), we are delighted that suspended sediments might be starting to be recognized as independent color-producing agents. This recognition bodes well for generation of value-added water quality products from space. Some valuable additions to the saga of case 2 water science may be found in Bowers et al. (1996); Gould, Jr., and Arnone (1997); Dekker et al. (1997; 2001b); Binding et al. (2003); Stramski et al. (2004); and Babin and Stramski (2004). Additional contributions to determination of numerical values of optical cross section spectra (particularly with respect to variations in scattering cross section spectra of geologically diverse suspended minerals) are hopefully forthcoming.

6.7 Back to the future with red wavelengths?

> Therefore, if direct measurements of subsurface irradiance reflectance, do, in fact, possess an application to the determination of water quality indicators, it is evident that the entire subsurface irradiance reflectance spectrum must be considered...however, single wavelength subsurface irradiance measurements in the red region of the visible spectrum do possess some applicability to the estimation of small to moderate concentrations of suspended minerals.
>
> From *Application of Direct Measurements of Optical Parameters to the Estimation of Lake Water Quality Indicators*, **Robert P. Bukata, J. Edward Bruton, and John H. Jerome, 1985**

Perhaps the preceding statement, which, at the time, we held to be self-evident, might warrant a caveat after more than 20 years. In the early days of Landsat-1, a profusion of misguided plots of ground-measured parameters vs. single or ratios of single wavelength bands emerged. Proponents of multivariate bio–geo-optical models have long advocated that full-color spectra be required to extract the coexltant concentrations of organic and inorganic CPAs from visible space radiation collected over inland and coastal waters. In theory, this remains an irrefutable fact. In practice, however, this has been a big-time obstruction to analyses of remotely acquired data (e.g., historical archives) that have lacked high spectral sensitivity.

A bane of satellite monitoring of natural water color (as discussed in Section 3.7) is the fact that almost every environmental entity strongly absorbs short wavelength (blue-end) visible radiation. The intervening atmosphere, along with each aquatic constituent (chlorophyll, suspended minerals, dissolved organic carbon), strongly absorbs downwelling blue light. (Absorption by molecular water, however, rapidly increases as the wavelength increases from blue to red.) Mathematically accounting for the removal of blue radiance values has often led to overaccounting, resulting in frequent determinations of erroneous estimates of negative water-leaving radiance values and subsequent erroneous estimates of negative chlorophyll concentrations. Thus, consistent with the need to provide products that may be of value to environmental end-users, Bukata et al. (2003) considered revisiting single-band (red wavelength) analyses.

Recall from Figure 3.1 and Figure 3.2 that (1) the water-leaving radiance spectra of chlorophyll-dominated case 1 waters are characterized by a distinct spectral peak in the red end of the spectrum (685 nm) due to fluorescence by algal pigments; and (2) such peaks are noticeably absent from water-leaving radiance spectra of non-chlorophyll-dominated case 2 waters. Although chlorophyll absorbs visible radiation, it scatters near-infrared radiation (the basis of NDVI). Thus, as chlorophyll concentration increases, the solar-stimulated fluorescence peak at 685 nm gradually evolves into one near 703 nm due to the interactive spectral consequences of absorption by molecular water and reflectance by chlorophyll — as Gitelson (1992), Gitelson et al. (1999), and others elegantly showed, using individual algal species and controlled ambient aquatic environments.

Scattering by chlorophyll of near-infrared wavelengths, along with the presence of nonzero concentrations of suspended minerals that scatter red wavelengths and attenuate fluorescing photons, also partly explains the lack of a distinct 685-nm fluorescence peak in most inland water volume reflectance, remote sensing reflectance, and water-leaving radiance spectra. It would appear, therefore, that remote sensing estimates of higher concentrations of chlorophyll would greatly benefit from a satellite band centered at 703 nm. Although these higher chlorophyll concentrations are beyond the spectral range of MODIS, MERIS does contain a spectral band to accommodate this shifted fluorescence peak.

Recapping, the absence of distinctive fluorescence peaks in most inland waters is partly explained by

- Relatively low (nondominating) concentrations of chlorophyll-bearing biota
- The shape of the 650- to 750-nm region of the subsurface volume reflectance spectrum in the absence of chlorophyll
- Near-infrared scattering by organic and inorganic particulates

- The presence of suspended sediments that scatter red wavelengths as well as attenuating downwelling fluorescence-inducing photons and upwelling fluoresced photons

Additional in-depth information may be found in Gower et al. (1999) and Brando and Dekker (2002).

Lack of a prominent fluorescence peak notwithstanding, as part of its CSA-supported GRIP project, NWRI is using in-water remote sensing reflectance as measured in Lake Erie with the WATERS device (suite of above-surface and below-surface spectrometers) to calculate the fluorescence line height (FLH) with wavelengths corresponding to MODIS bands. Figure 6.2 (taken from Figure 5 of Bukata et al., 2003) shows a somewhat unexpected, moderately good relationship ($r^2 = 0.66$) between directly measured chlorophyll concentration and FLH as determined from WATERS spectra.

Such curves (irrespective of possibly indifferent statistics) illustrate that an absence of a well-defined spectral fluorescence peak is not necessarily indicative of an absence of chlorophyll fluorescence in case 2 waters dominated by CPA other than chlorophyll. They also serve to reinforce the *Gedanken* thought process discussion of Section 3.5 dealing with the need to determine *in situ* (as opposed to laboratory) values of indigenous CPA optical cross section spectra because water color, whether recorded directly or remotely, is a consequence of *all* the photonic interactions (elastic and inelastic) occurring within the water column. As seen in Chapter Five,

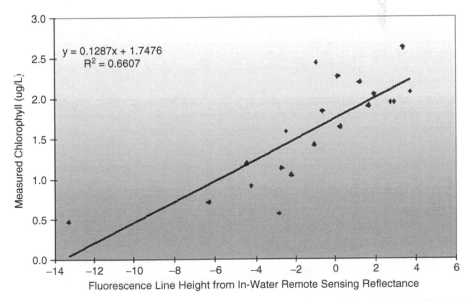

Figure 6.2 Relationship between directly measured chlorophyll concentration and fluorescence line height as determined from WATERS. (From Bukata, R.P. et al., *Backscatter*, winter issue, 29–31, 2003.)

water-leaving radiances in the red region of the visible spectrum have practical applications to sediment transport phenomena and other general inorganic turbidity concerns of environmental managers, stewards, and policy-makers. Curves such as that shown in Figure 6.2 illustrate that an understanding of inland water fluorescence can result in the ability to use water-leaving radiance at red wavelengths to estimate concentrations of aquatic chlorophyll also.

The need for hyperspectral remote sensing and the development and application of robust multivariate bio–geo-optical models to monitor inland water quality reliably is inarguable. However, the fact that a simple and restricted spectral region (specifically, red wavelengths) can be a valuable component of inland water quality monitoring should not be lost. Two of the most common water quality parameters measured by aquatic biologists are *turbidity* and *chlorophyll*. Although obvious lack of scientific precision and composite inclusiveness of each of these terms is acknowledged, they do serve as meaningful fiducials for a variety of environmental management purposes. Satellite, airborne, and in-water determinations of stimulated solar fluorescence and water color have shown that turbidity and chlorophyll products can be provided by a simple system with a few spectral bands in the red wavelength region of the visible spectrum.

Furthermore, by restricting satellite band analyses to the visible red and short near-infrared (due to the 703-nm shift) region of the water-leaving radiance spectrum, the problem of negative blue radiance values evaporates and a very simple or possibly even no atmospheric correction algorithm would be required to produce these simple information products (although the latter may be a dream). The combination of a few bands and less atmospheric correction uncertainty (at least for the so-called clear days) greatly reduces the complexity and costs of producing these simple products.

Borstad et al. (2001) proposed that a red wavelength-focused piggy-back sensor termed the "Red Nymph" mounted on the palette aboard the International Space Station could provide water quality managers and environmental policy-makers with near real-time turbidity and chlorophyll information across the inland and coastal waters over the highly populated areas along the Canada–U.S. border. Although the conceptual Red Nymph satellite sensor in no way invalidates the need for hyperspectral remote sensing sensors (for research and for applications of that research), the simple aquatic information products (turbidity and chlorophyll) are important to water quality stewards. Furthermore, such simple information products have counterparts in other areas of environmental importance. Data from the same few red and near-infrared wavelength bands can be used to provide vegetation vigor and biomass information for forestry, agricultural, and other land-use assessments.

One principal concern of watershed and coastal zone managers is the ongoing threat of aquatic eutrophication, which may lead to nuisance or toxic algal blooms, anoxia, fish-kill, and other disruptions to land–water dynamic equilibria within the watershed or coastal zone. Eutrophication is generally the result of excessive aquatic nutrients entering the lake or coastal zone as run-off from land-use activity such as industry, forestry, farming, sewage disposal, and recreation. Clearly, the simple red-band analyses (with a bit of encroachment into the near-infrared) can produce water quality products valuable to watershed and coastal zone management. These analyses can be provided in a manner that, although not satisfying the doctrine of *science über alles*, does, if cleverly applied, satisfy the principle of science/ use of science compromise. The suggestion of a Red Nymph satellite sensor is definitely worth pursuing.

Although we must continually look forward, we should not forget that some clever and creative thought and research are behind us and that an occasional look in the rear-view mirror might be of great consequence. Many things old may become new once more.

> Of this I am quite sure, that if we open a quarrel between the past and the present, we shall find that we have lost the future.

Sir Winston Churchill, June 18, 1940

chapter seven

Truth in advertising of remote sensing products

Men occasionally stumble over the truth, but most of them pick themselves up and hurry off as if nothing has happened.

Sir Winston Churchill, circa 1935

A half truth is a whole lie.

Old Yiddish proverb

The probable real and perceived root causes for the persistent malaise that has plagued remote sensing, in general, and remote sensing of inland waters in particular include, among others:

- The inherent irony of remote sensing
- Generators of remote sensing products as the principal end-users of their products
- Financial inaccessibility of archival data
- Technology leading science leading applications
- Lack of public awareness
- High cost of deliverables
- Lack of official standards
- Ecosystem synergisms confounding mathematical relationships
- Calibration and validation of bio–geo-optical models and algorithms
- Intractable atmospheric aerosols
- Unswerving, dedicated adherence to historical environmental monitoring protocols
- Dubious marketing practices

  • Proliferation of computer programs that remove modelers from
    the remote sensing firmament
  • Organization barriers

  However, the most damaging cause is the suspect reputation
arising from early unsubstantiated and unfulfilled promises that
defined the pedigree, and thereby became the mantra, of remote
sensing. This sad mantra emerged from the 1960s social, moral,
and political doctrines of "all inclusiveness except for elitism"
and ensuing re-evaluations of roles that science should play within those doctrines. Although our crystal ball gazing of Chapter
Six seemed to show the future of monitoring Earth's environment
from space to be more promising than its past would have led us
to hope, the crystal ball itself was not free from clouds.

7.1   *Environmental science and the vagaries of populism*

The history of environmental science and the vagaries of populism continue
to be an interactive potpourri of fact, myth, real and perceived environmental
threats, environmental policy, science, and extant public perception of the
role of science linked together by contradictory rhetoric. Science after the
Renaissance advanced through curiosity, creativity, deduction, and experimentation. Theory, instrument design, laboratory experimentation, field
excursions, data analyses, and model verification were the generally perceived roles of science. In developed countries, science was considered a
noble, nonlay entity. Although not actively divorced from public scrutiny, it
was generally not under a public microscope. Science was advancing by
intellectual elitism rather than societal dictates.

  The social unrest, all-inclusiveness dogma, and real and perceived
threats to Earth's environment of the 1960s produced an upheaval in populist
visions of science. Theoretical research was demonized as self-centered pas-
times of garret-sequestered old men out of touch with everyone but other
garret-sequestered old men; *mission-oriented research* emerged as the new role
for science as perceived by populists. Academic institutes, government funding agencies, and the private sector scrambled to define "mission" in a
manner that would satisfy the disparate dictates of a nonscientifically trained
populace. Therefore, environmental science and remote sensing pursued
some ill-defined, all-inclusive, eco–social missions in order to generate some
ill-defined, all-inclusive, eco–social products.

  Undeniably, scientific research should contribute to public benefit, and
historically (although at times controversially) it generally has. The first half
of the 20th century was largely served by physics research (classical, atomic,
nuclear, solid state, space) and the second half largely by chemistry, biology,
and bio–genetics research. Fortunately, medical research served throughout.

However, a free-world population preoccupied with the advancement of individuals and organized special interest groups often becomes far more influenced by political ambitions cloaked in righteous visions than it does by factual science.

In such manner, the turbulent decades of the 1960s and 1970s witnessed the surreptitious removal of considerable deductive reasoning (the fundamental basis of science) from the realms of trained dispassionate analysts and planted within the newly established realms of passionate political and social activists. These passionate activists (assembled from movie lots, concert stages, sports arenas, coffee houses, liberal arts colleges, news media, and nongovernment organizations) steadily assumed unchallenged postures as self-appointed spokespersons in topics (such as civil rights, social equality, justice, liberty, world peace, and Earth's environment) that involved visions of alternative populist realities. Many of these visions (an end to bigotry, wars, world hunger, persecution, apartheid, racial genocide, pollution, and spread of AIDS) were of such superficial morality and nobility that rational scientific debate was not only difficult to initiate but also futile.

One is hard pressed to debate apparent nobility with pop-cultural icons bent on influencing general public mind-sets. Therefore, publicly perceived, righteous proenvironment crusades (ban pulp- and paper-bleaching processes, prohibit culling of baby harp-seals, initiate shark stock-rebuilding programs, ban use of fertilizers, and create artificial nesting grounds for migratory birds) can stymie necessary scientific debate. This occurs when

- Impartial science is supplanted by the politically correct "selective science" of crusading visionaries, i.e., justification predicated upon observations or measurements that, irrespective of context, are used to advance opinion rather than fact.
- Any impartial science that challenges the nobility (and, thus, validity) of such selective science is promptly discounted by the media as a mean-spirited, environmentally threatening fallacy designed to preserve the status quo currently under siege.

In this manner, some founded fears of *local* and many unfounded fears of *global* anthropogenic disruption of ecosystem stabilities in Earth's biosphere created a guilt-ridden, environmentally conscious populace that dutifully recycled, composted, rode bicycles, organically grew vegetables, gave up air conditioning, stopped killing mosquitoes, and sought to ban anything chemical. Endorsed by pop-culture entertainers, left-wing news media, liberal arts colleges, and a continuous catalogue of selective science, passionate promoters of environmental policies and visions were secure in their bunkers atop the moral high ground. They could usually deflect criticisms of the fear-instilling prophecies put forth by crusading environmental activists, thereby "proving" the wisdom of the policies to the general public. All too commonly, crusading visionaries on the "pro" side of fervent beliefs (activists, talk-show hosts, television comedians, rock music bands, movie screen icons, athletes, politicians, singers, lawyers, union leaders) are active in the media; however, those on the "con" side of those fervent beliefs (despite agreeing with the nobility of the vision) do not crusade and are inactive in the media.

As a consequence of its origin in the 1960s' confounding admixture of fact, myth, real and perceived environmental threat, and extant public perception of the roles of science, space monitoring of continental environments was launched on a slippery slope: it had been denied sufficient dispassionate science, which it lacked but required to be considered a legitimate tool for environmental assessment. Fortunately, due to the established and acknowledged importance of oceans to Earth's climate, space monitoring of ocean environments was not.

Sadly, the evolution of environmental science and the vagaries of populism have not changed significantly from the course set in the turbulent 1960s. Numerous preordained environmental policies are driven by preordained fervent beliefs. Dispassionate science is and assuredly *should be* a powerful force. As such, science-based products emergent from space monitoring of environmental change in general (and inland and coastal water quality in particular) must not be perceived as supporting unfounded fervor.

7.2 Three examples of controversial environmental issues

Judgmentally selective science is often an unwitting pawn for well-meaning and self-serving agendas of environmental activists, who are not necessarily scientifically trained. Assuming the role of a devil's advocate, let us consider three examples of current controversial environmental issues that are more under sociopolitical and liberal arts control than dispassionate scientific guidance. Let us also examine a few "con" sides to those issues that, when presented, are invariably ridiculed and dismissed as irrelevant by the media-supported crusading advocates of the "pro" sides.

7.2.1 Climate change (a.k.a. global warming)

In Section 7.1, we have lamented that environmental activists and public media may manipulate diverse and unrelated measurements and observations to promote steadfast opinions:

- The United Nations regularly produces documents that attest to its unrelenting belief in global warming disasters.
- In his best-selling book, Al Gore (1992) predicts imminent irreversible environmental damage from global warming if industrial outputs are not immediately curtailed and automobiles banned.
- Apparently declaring the scientific controversy over, *National Geographic* (September, 2004) devoted 74 pages to "proving" its proclamation that Earth is definitely and rapidly getting hotter and that

the only important issue is our unwillingness to slow the meltdown by curbing our use of fossil fuels.

- Hollywood has promoted its support of global warming devastation with the opulent special-effects doomsday thriller, *The Day after Tomorrow* — evidently patterned after the destruction of Superman's fictional home planet, Krypton.

By contrast, however, very little public media and surprisingly little mainstream scientific journal coverage has been given to such counterarguments as

- The natural solar-induced climate change that has caused the Earth to progress through three major and several minor devastating ice ages (between each of which global warming obviously occurred)
- The minuscule role (Section 1.2) played by anthropogenic carbon dioxide as a "greenhouse gas" (industrial CO_2 is a nontoxic by-product to be abated, but its role as an agent of global warming is scientifically in question)
- Supportive evidence that, despite the absence of an Industrial Revolution, the inferred total atmospheric CO_2 concentrations between major Ice Ages 2 and 3 are considerably higher than the inferred total atmospheric CO_2 concentrations of today
- Images from the Mars orbiter camera aboard Mars Global Surveyor that have elegantly revealed that the Martian polar ice caps have experienced markedly different geologic pasts and that yearly seasonal spring defrosting and mass movement of ice cap material result from large changes in the year-to-year heat budgets of polar cap regions
- The severe limitations of GCMs in defining *present* climate regimes, let alone predicting *future* climate regimes
- The inherent and logistical difficulties and uncertainties in estimating global temperature and total atmospheric carbon compound content
- The lack of direct evidence that pollution and climate change are interchangeable as cause–effect entities
- The skewed predominant placements of temperature gauges within or close to major cities where increasing populations and asphalt surfaces generate increasingly warmer ground-level thermal hoods
- The fact that satellite thermal sensors have not observed temperature increase in the upper atmosphere
- The fact that the same temperature data sets used initially by earlier environmental activists to forewarn of doom by global *cooling* are now used (much more popularly) by current environmental activists to predict doom by global warming (*The Day after Tomorrow* belies this apparent contradiction by a plot premise that global warming rapidly leads to an Ice Age.)

Considerably more expanded devil's advocacy may be found in Michaels and Knappenberger (1996); Bate (1996); Hoyle (1996), Boehmer–Christiansen (1996); Doran et al. (2002); Kininmonth (2004); and elsewhere. Perhaps the most reasoned, in-depth, scholarly analyses of the politics–science–climate-change issue have been presented by S. Fred Singer. Among his numerous other works, see Singer (1997, 1999).

Since 1998 the central piece of "scientific evidence" continually presented by the UN, other proponents of anthropogenic-induced global warming, and advocates of the Kyoto Protocol has been the so-called "hockey stick" image of a 600 year time plot of Northern Hemisphere temperatures that is essentially stable for five centuries before turning sharply upwards at the 20th century. The "hockey stick" papers were initially published by Michael Mann et al. in 1998 and 1999. As this book was being prepared for publication, a research paper authored by Stephen McIntyre and Ross McKitrick (undoubtedly destined to ignite much-needed scientific debate) has been accepted for publicaton by the same scientific journal (*Geophysical Research Letters*) that published the 1999 Mann et al. calculations. The 2005 paper criticizes the statistical approach used to devise the "hockey stick" image as being flawed and statistically meaningless. The steadfast belief in anthropogenic global warming within the academic community and the news media has already dismissed the McIntyre-McKitrick counter argument as wrong and irresponsible. Controversy will certainly ensue. The editors of *Geophysical Research Letters* are to be congratulated for upholding the scientific principle. Scientific papers referred to here are:

> Mann, M.E., Bradley, R.S., and Hughes, M.K., Global-scale temperature patterns and climate forcing over the past six centuries, *Nature*, 392, 779–787, 1998.
>
> Mann, M.E., Bradley, R.S., and Hughes, M.K., Northern Hemisphere temperatures during the past millennium: inferences, uncertainties, and limitations, *Geophys. Res. Let.*, 26, 759–762, 1999.
>
> McIntyre, S. and McKitrick, R., Hockey sticks, principal components, and spurious significance, to appear in *Geophys. Res. Let.*, 2005.

7.2.2   *Depletion of the stratospheric ozone layer, and the ensuing environmental havoc*

The Earth's thin layer of atmospheric ozone protects life on Earth by absorbing biologically pernicious solar ultraviolet radiation (all UV-C, most UV-B, but no UV-A). The launch of NASA's prototype total ozone mapping spectrometer (TOMS) in 1978 on Nimbus-7 followed by the launch of its successor in 1996 on Earth Probe has enabled high-resolution measurements of columnar atmospheric ozone concentrations from space to be displayed as daily maps of atmospheric ozone distribution. These TOMS global ozone distribution maps are readily available on the Internet.

TOMS delineated the inappropriately called "Antarctic ozone hole" (well-defined, large-scale, recurrent natural destructive thinning of the ozone concentration over Antarctica to <220 Dobson units, which has occurred every Antarctic spring with the possible exception of 1956). These delineations, with concurrent delineations of dramatic ozone variations, including initially well-defined progressive ozone thinning in northern polar latitudes, understandably resulted in concern that concomitant increased levels of UV-B would reach Earth and dramatically affect human and ecosystem health. Therefore, UV-B was declared an urgent environmental threat and government, academic, and public media teams were mobilized to compile evidence of UV-induced public health risk and environmental distress. The "science in support of public policy" agenda began to be driven by fears of

- Melanoma
- Immune system suppression
- Fish eye cataracts
- Loss of wetland and aquatic habitats
- Arctic caribou blindness
- Amphibian death and disfigurement
- Reduced ocean productivity
- Genetic mutations
- Massive migrations of fish, mammals, and vegetation
- Disruptions to ambient food chains
- Deterioration of terrestrial landscape infrastructure

Numerous other perceived threats arose, as well as their social, political, and economic consequences. Once again the "future shock" alarm was sounded and once again "guilt by anthropogenic destruction" was the unanimous verdict. In this instance, hydrochlorofluorocarbons (HCFCs) were the indicted culprits and U.N. treaties were quickly signed to prohibit their industrial and public release. Scientific counterarguments that are still regarded as minority views include that

- Earth's ozone layer is in a state of dynamic equilibrium in which atmospheric ozone is continuously being naturally produced and destroyed by solar-induced photoreactions.
- Atmospheric ozone is unequally distributed about the Earth in a manner governed by combined orbital motion of the Earth around the Sun, diurnal rotation of the Earth about its own axis, and general circulation patterns present within the atmosphere.
- The inevitable results of these dynamic constraints are that ozone minima occur over the equator and maxima over northern hemisphere mid-latitudes and southern hemisphere Antarctic, and that the tilt of the Earth's axis to the plane of the ecliptic results in spring seasonal decreases in stratospheric ozone in northern and southern high- to mid-latitudes.

- Solar activity follows the long understood 11-year solar cycle which, in general, manifests as a 3- to 4-year rise to its maximum activity followed by a 7- to 8-year decline to its minimum; no two cycles are identical and the density of stratospheric ozone would also follow this 11-year cycle.
- Earth Probe TOMS, the catalyst for much of the furor over ozone "holes," was launched on the declining cycle of solar activity; several years after the U.N. protocols were in place, the solar cycle was near its maximum and the observed "hole closing" was conveniently attributed to enforced removal of HCFCs.
- HCFCs, which are many orders of magnitude denser than air, would find it difficult getting to the stratosphere; most would fall to the ground (agricultural studies have indicated large in-soil concentrations of HCFCs that, however, do not inhibit plant growth) beside their emission site. The very few industrial HCFCs that *might* become airborne would benefit the public by attacking the low-level anthropogenically generated "bad" ozone that constitutes the health threat known as smog.

7.2.3   Attacks on aquaculture practices as being detrimental to wild fish populations

The immediate and urgent issue of rapidly depleting global stocks of food-fish has resulted in rapid growth and expansion of (1) aquaculture (cultured pond-farms broadly classified into macroalgal and/or microalgal culture, shellfish culture, crustacean and/or fish pond culture, and fish cage culture); and (2) wild fisheries (wild fish pond-farms managed using stock assessments based on single species population models with direct focus on a collapse or a risk of collapse of a high-profile fish stock). Most of these aquaculture/wild fishery activities are conducted in designated coastal areas where, not unexpectedly, complexities arise from natural and cultivated ecosystems sharing the same ocean.

We will not presume to present a dissertation on these well-documented issues here. Suffice to say that these issues include:

- High levels of pond eutrophy resulting from added feed and *in situ* bioproduction
- Moderate to high exchange rates between pond-farms and adjacent waters
- Maintenance of high-quality environments for fry and adult fish
- Recycled nutrients in sheltered waters of low flushing rates
- Vulnerability of shellfish and caged fish to harmful algal blooms
- Development of benthic anoxia

Many others issues fall into this general category of biological responses to simultaneous exposure to case 1 and case 2 waters.

Aquaculture and wild fisheries are playing welcome and essential roles in addressing (with careful attention to the complexities) a very serious environmental issue by restoring populations of valued fish and seafood species. However, a cadre of nonscientifically trained lobbyists who are against farmed fish, under aegis of privately funded organized societies with proenvironment names and purposes, have accused managers of cultured and wild fish/seafood farms of *possibly* infecting noncultivated fish stocks (wild pink, chum, and sockeye salmon are high on the list) with lice and disease. Although far from a universal condemnation of fish/seafood culturing, such an accusation does describe probable conditions that, if not controlled, become serious. However, wild fish can and have encountered lice and other ills far-removed from fish farming.

Other accusations claim that farmed fish (salmon in particular) contained higher levels of PCBs than their wild ocean counterparts. It is our current understanding that (1) scientific debate as to whether this PCB accusation is true is ongoing; and (2) whether or not the accusation is true, *both* natural and cultivated salmon stocks have PCB levels lower than those in most other foods and are well within North American consumptive food regulations. The inherent fear-mongering accompanying the nonscientifically supported accusations resulted in news media and some global agencies endorsing recommendations of those lobbying against farmed fish as politically correct. These included banning consumption of cultured salmon in favor of consumption of the "healthier" wild salmon (thus far optional).

An environmentally conscious public quickly responded. Purchase and consumption of canned wild salmon has escalated. Environmental activists' advocating increases in the catch and the consumption of a vanishing fish stock (i.e., saving an endangered species by eating it) would certainly appear inconsistent with and less well thought out than many of their previous proselytisms. Would these same environmental activists preach that to save endangered forests we should accelerate harvest of the mature trees rather than planting new ones?

7.2.4 Subsection summary

Inarguably, the environment is threatened by pollution, overpopulation, and overharvesting of its natural resources. Protecting and rehabilitating the environment is essential for the future of the planet. Also inarguably, climate change is a very real, natural, nonstatic, and ongoing process. It is easy to concede that natural climate change and natural environment change are related. It is also easy to concede that local anthropogenic pollution and local (and, in many cases, somewhat distant) *environment* change are related. Through direct injection at a fixed site or subsequent to its atmospheric or aquatic transport to a distant site, pollution does have an adverse impact on an otherwise undisturbed environment.

However, it is less easy to concede that anthropogenic pollution and global *climate* change are related. Establishing sound practices and policies based on environmental protection from toxicants, refurbishment of exploited resources, and adaptation to and/or mitigation of stresses of changing climate is appropriate, essential, and laudable. It is obvious that science must be the cornerstone to practices and policies established and enforced to ensure such protection, refurbishment, adaptation, and mitigation. For this reason, a "science in support of policy" dictum was born and quickly adopted as politically expedient and politically correct. However, such a dictum is suggestive of a "cart-before-horse" syndrome implying an unworkable role reversal. Perhaps, "policy in support of science" would be a more logical dictum because a policy surely should emerge *subsequent to* supportive definitive science, rather than "definitive" science being required to support a proposed or already established policy.

It is politically astute and acceptable to an already guilt-ridden and environmentally conscious public to brand anthropogenic carbon compounds as agents of global warming through a greenhouse effect. Indeed, carbon dioxide and the global environment have been and are historically related. However, within this historical relationship, another role reversal is suggested if carbon dioxide is considered a *cause* and global climate change is considered an *effect*. By far, the greatest injections of carbon dioxide into the Earth's atmosphere (far greater than those injected yearly from fueled transportation and industrial activities) are from volcanic eruptions driven by thermodynamic activity within Earth's core. This would suggest that climate change is a cause and atmospheric carbon an effect.

Life on Earth is carbon based and depends on vital, mutually supportive animal, plant, oxygen, carbon dioxide, inhalation, and exhalation relationships. Forests and oceans are effective sinks for atmospheric carbon, despite some lobbyists' convictions that they have become sources of second-hand smoke. Of course, these and other environmental realities neither imply nor suggest that enforced policies regulating the injection of anthropogenic carbon compounds into the local environment (along with the development of home and business energy efficiencies, clean-burning coal, fuel-cell technologies, and alternate energy sources) are misguided, unnecessary, or wrong. They might imply, however, that justification of such policies and activities as mitigating greenhouse gas-induced global climate change are.

7.3   *Quality science emergent from fear-instilling rhetoric*

Memberships in nongovernment, proenvironmental organizations, as with many vacancies in science-based news media, are filled with more emphasis placed on media know-how than on scientific or technical expertise. Actual public good often becomes subordinate to fear-instilling rhetoric to attract publicity and raise academia's government research funds and/or donations from special interest groups. However, on occasion such rhetoric does produce some quality science. Two such occasions resulted in well-documented dramatic reductions of water quality impacts from eutrophication and "acid rain" generated by anthropogenic insertions of industrial and residential pollutants into the biosphere.

In truth, however, the quality science of the eutrophy and atmospheric acidic deposition issues emerged from a combination of theoretical modeling and ground-based research and observations. Remote sensing was most often absent; when present, its attempts to provide quality science were misleading or wrong. During the height of scientific and public concern over acid rain (early 1980s onward), many proposals were written and many funded to apply existing methodologies for utilizing remotely sensed water color to determine acidity of inland waters. Unfortunately, "existing methodologies" widely published by several authors and quoted within those proposals were based on two improbable myths:

- Chlorophyll concentrations in inland waters are unambiguously related to pH (pH = 7 represents neutrality; pH < 7 represents acidity; pH > 7 represents alkalinity).
- Chromaticity analyses (apportionment of white light into its red, green, and blue components) can inversely model unique values of chlorophyll concentrations from inland water color.

Using their multivariate optical models and data from over 200 lakes in the highly affected areas around Sudbury, Ontario, Bukata et al. (1982, 1983) dispelled both these myths and enabled the EC-led Long-Range Transport of Atmospheric Pollutants (LRTAP) Program to avoid the useless expenditure of hundreds of thousands of dollars in contract funds. Nonetheless, the struggling reputation of environmental remote sensing was further demeaned by such endorsed myths, as were the products of remote sensing vendors. In a University of Sherbrooke, Quebec, bibliography of remote sensing publications, the previously cited 1982 paper carries the impertinent, asterisked citation: "a refreshing slap in the face to remote sensing marketing types." Hopefully, vendors of remotely sensed inland water quality products are now well beyond requiring face slaps.

The sharp focus in the mid-1990s on possible anthropogenic diminution of the stratospheric ozone layer presented another occasion in which quality science emerged from fear-instilling rhetoric. In this instance, however, remote sensing did not fare as badly. UV-B is well known as a hazard to human, plant, and wildlife — a cause of sunburn, eye cataracts, and skin cancer and a threat to ecosystem (land, water, wetland) stability. Apart from the mobilization of quickly assembled teams of academics to board the bandwagon about to be funded, the potential threat of enhanced levels of ground-based ultraviolet radiation did reveal several gaps in scientific knowledge. Two major gaps that were addressed were the limited information on (1) the action spectra (responses to an incident irradiance spectrum) of a variety of organisms and higher life-forms to ambient, reduced, and enhanced levels of ultraviolet radiation (UV-A, UV-B, UV-C); and (2) the dispersal and associated synergistic consequences of ultraviolet radiation within an environmental ecosystem.

Ultraviolet radiation is the energy source for many photoprocesses occurring within Earth's environment. Some processes are photochemical and may result in production or destruction of various chemical species at molecular, atomic, or ionic levels. Other processes are photobiological and may result in damage to an organism at the molecular or cellular level, thereby inhibiting the organism's growth, disrupting its functionality, or even causing its death. The contribution of energy at each ultraviolet wavelength to a photoprocess is determined by the action spectrum specific to that photoprocess. TOMS observations underlined a crucial need to quantify (1) the photobiological and photochemical processes of greatest risk to ozone depletion; (2) the critical level of ozone depletion for these photoprocesses; and (3) an intercomparison of responses of diverse action spectra to changing ultraviolet radiation.

Within a few years, a wealth of literature appeared (some excellent papers with hitherto unknown scientific intelligence, some speculative with no scientific credibility, and some with special-interest and self-serving agendas). Workshops focusing on science, issues, impacts, policies, and programs involving atmospheric ozone, UV-B, and climate abounded. Academic, government, and private sector researchers, environmental managers, and policy-makers considered a wide variety of interactive topics, including total atmospheric ozone; human health and vision; agriculture; terrestrial and aquatic ecosystems; fishing; infrastructure; and disruptions to socioeconomy, native people customs, and city, suburban, and rural lifestyles.

7.4 Ground-level ultraviolet radiation and natural waters

With respect to aquatic systems, impacts of enhanced levels of UV-B to marine ecosystems were more established and well understood than for inland waters — as had been the earlier situation for oceanographic remote sensing. Valuable information may be found in Harm (1980); Calkins and Barcelo (1982); Worrest (1983); Smith and Baker (1989); Ekelund (1990); Häder and Worrest (1991); Cullen and Lesser (1991); Cullen et al. (1992); and Kirk (1994b). Anthologies such as those edited by Calkins (1982), Williamson and Zangarese (1994), Häder (1997), and de Mora et al. (2000) also offer important information.

Concern regarding potentially disappearing concentrations of upper atmosphere ozone and possible impacts to freshwater ecosystems resulted in the need to fill gaps in knowledge of the bio–geo–chemico–physical complexities of inland waters. In addition to the need to determine action spectra of a variety of aquatic components, it was also necessary to understand the synergisms and impacts associated with ecosystem responses to multiple environmental stressors. Research focused on chemical, biological, and human health studies was quickly established globally at numerous government, university, and private institutes. Clearly, UV radiative transfer mechanisms in inland waters were integral to such understanding. Thus, many of the techniques successful in development of bio–geo-optical models for visible and near-infrared wavelengths could be logically extended to ultraviolet wavelengths. As an illustration of such optical research, we will briefly discuss some of activities initiated at EC's National Water Research Institute.

An interesting perspective on general ecosystem response to interactive multiple stressors was presented by Bukata and Jerome (1997). These scientists used distributions of phytoplankton biomass in monitored, controlled, and replicated enclosures in the shallow pelagic region of a boreal lake (that were then separated from external waters) to expand studies of UV-B radiation impacts to aquatic population dynamics beyond well-documented algal colonization and herbivore predation. Subsequent to their being embedded into the lake-bed sediments, selective enclosures were exposed to ambient UV-B radiation (open to sunlight), removal of UV-B radiation (solar filter), or enhanced UV-B radiation (artificial source).

Disparate evolution of phytoplankton populations among the three enclosure types, coupled with surprisingly similar evolutions within each replicate, led to the tempting musing that aquatic ecosystems (at least at the lower trophic levels) that evolved under ambient (i.e., "normal") environmental conditions possess optional degrees of freedom (and, consequently, optional futures) that are denied ecosystems subjected to unexpected stress (i.e., departures from normalcy due to increase, decrease, or total removal of a synergistic variable constituting part of the ambient environment, or the insertion of a synergistic variable not originally part of the ambient environment). The restricted evolutionary protocols invoked by the removal of possible futures appear to be (1) directed towards combating or adjusting to this departure from normalcy; and (2) based upon an extant "survival of the fittest" scenario.

UV-B radiation incident at the air–water interface is highly diffuse and results in an essentially constant aquatic surface reflectance of 6 to 8% independent of solar zenith angle. UV-B is also highly attenuated by dissolved organic matter (yellow substances such as organic carbon). As discussed and illustrated in Bukata et al. (1995), the depth of the 1% subsurface irradiance level for the entire UV-B spectral range (280 to 320 nm) decreases by about two orders of magnitude as the water column concentration of DOC increases from zero (where the 1% irradiance level is at depth ~40 m) to 5.0 mgC L^{-1} (where the 1% irradiance level is at ~0.3 m). For a DOC concentration of 2.0 mgC L^{-1} (average Lake Ontario value), 99% of incident UV-B is absorbed in the upper 1 m (or, equivalently, 90% is absorbed in the upper 0.5 m). Considerably higher DOC values are found globally in a significant number of inland and coastal waters.

Thus, a decrease in atmospheric ozone with its manifest increase in ground-level UV-B irradiance would be a more serious problem to oceanographic ecosystem stability (e.g., marine producers in the Antarctic and other high-latitude ecosystems) than to most inland and coastal aquatic ecosystem stability. This was illustrated in the work of Jerome and Bukata (1998a) in which they tracked solar extraterrestrial ultraviolet radiation from its impingement upon the Earth's atmosphere to its ultimate dispersal in natural inland waters. To do so they utilized:

- The combined optical models of Kirk (1984), Bukata et al. (1985), Jerome et al. (1988), and Morel (1988)
- The extraterrestrial radiation spectra of Heath et al. (1975)
- The atmospheric UV-B irradiance models of Green et al. (1980) and Green and Schippnick (1982)
- The biological weighting functions of Cullen et al. (1992), Villafane et al. (1995), and Setlow (1974)
- The optical cross section spectra from several inland water bodies
- Directly measured (from the EC-MSC monitoring network) values of ground-level ultraviolet radiation and stratospheric ozone.

Subsequent to its downwelling through the atmosphere and through the air–water interface, solar UV-A and UV-B are dispersed by and within the aquatic medium. This dispersal defines the underwater UV photon budget as the final apportionment of the UV-A and UV-B photons that have reached the air–water interface. Figure 7.1, taken from Figure 15 of Jerome and Bukata (1998a), illustrates (and is representative of other inland water bodies) the UV photon budget as a function of wavelength (290 to 400 nm) for a Lake Ontario water column devoid of suspended sediment but containing 10 µg L^{-1} chlorophyll a and 2.2 mgC L^{-1} of DOC. The salient features of Figure 7.1 are:

- Comparably small (~5%) and relatively spectrally invariant fractions of the UV photons are reflected by the water surface and absorbed by chemically pure water molecules. The remaining 90% of the UV-A and UV-B photons are dispersed between the indigenous columnar chlorophyll and DOC.
- Two distinct peaks (~25% at $\lambda \sim 338$ nm and >40% at $\lambda > 400$ nm) define the maximum chlorophyll contribution to the UV photon budget. A minimum chlorophyll contribution (~13%) to the UV photon budget occurs at ~360 nm. At the shortest wavelengths, the chlorophyll contribution to the subsurface UV photon budget is comparable to that of the pure water.
- The contribution to the subsurface UV photon budget by the DOC is in anti-phase with the contribution of chlorophyll (minima of ~65% at ~338 nm and <55% beyond 400 nm). Maximum DOC contributions to the subsurface UV photon budget are displayed at the shortest wavelengths (~86%) and at ~360 nm (~80%). Clearly, this preferential (close to total) absorption of the biologically deleterious UV-B photons by the DOC emphatically illustrates the protection provided to

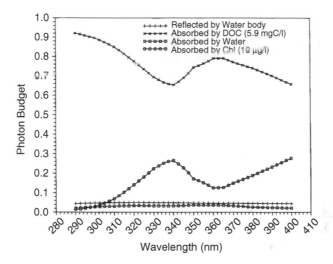

Figure 7.1 Ultraviolet photon budget as a function of wavelength (280 to 400 nm) for a Lake Ontario water column containing 10 µg L^{-1} chlorophyll *a* and 2.2 mgC L^{-1} of DOC. (From Jerome, J.H. and Bukata, R.P., *J. Great Lakes Res.*, 24, 666–680, 1998a.)

aquatic organisms by the presence (even in relatively small quantity) of dissolved organic matter in the water column. Conversely, it also indicates the vulnerability of aquatic organisms (for example, in pelagic ocean regions where the 1% UV-B irradiance depth is ~ 40 m) in the absence of columnar dissolved organic matter.

Throughout the entire UV-A and UV-B spectrum, the resident DOC accounts by far for the largest component of the UV photon budget. Between ~86% (at 290 nm) and ~76% (at 320 nm) of the UV-B photons are absorbed by the DOC. The invariable presence of suspended inorganic matter (considered absent in the UV dispersal scheme of Figure 7.1) affords further protection to aquatic organisms. As visible wavelengths are approached, the UV photon dominance by DOC gives way to PAR photon dominance by chlorophyll *a* (at λ = 400 nm, the contribution of chlorophyll has increased to ~35% and rising, and the DOC contribution has decreased to ~55% and falling).

In a comparable study, Jerome and Bukata (1998b) modeled the clear sky ground-level UV irradiance spectrum over Lake Ontario on the vernal/autumnal equinox and the summer solstice for two concentrations of stratospheric ozone. Using these UV spectra with appropriate apparent quantum yield spectra diffused rising fears that the worst case scenarios of deteriorating stratospheric ozone concentrations would result in serious photoreproduction of hydrogen peroxide in the lower Great Lakes.

Although much of the hype has subsided and many of the exaggerated threats and fears associated with early TOMS maps of expanding "ozone holes" have been proven groundless, the hard science that emerged in a short space of time harks back to the scientifically productive era of space science and space exploration. It also serves to illustrate that remote sensing (despite continued presence of special-interest manipulating) can, with assistance of well thought out research, approach a controversial issue with (and preserve) scientific credibility. This scientifically credible approach must be harnessed, nurtured, expanded, and unleashed on products for public good resulting from space applications such as discussed in Chapter Five.

7.5 Environmental monitoring from space and controversy

We have sketched three (not an exhaustive number) highly controversial, volatile topics in which political, scientific, social, and public media ideologies are in conflict. For these unsettled controversies, neither compelling nor convincing scientific evidence has been presented to warrant the political and social fervor aroused. International environmental treaties that have had an impact or soon will have an impact on lifestyle, custom, economy, and progress have been signed, or not been signed, with little, no, or insufficient scientific evidence to support either decision.

One principal, publicly perceived role of terrestrial and aquatic monitoring from space is recording and analyzing weather- and climate-related changes (Chapter Four and elsewhere). Such a role is also perceived as integral (and in many instances paramount) to the active and planned national and international EOS programs discussed in Section 5.1. As such, space monitoring becomes central to the currently major environmental controversy. Climate-change controversy over inland and coastal water *quality* products is probably less of a potential problem (because run-off toxicants are not CPAs) than climate-change controversy over water *quantity* products (floods, advance and retreat of glaciers, tsunamis, shore processes). However, satellite monitoring cannot ignore consequences of manipulated public emotion possibly becoming a nemesis of scientific dispassion. Historical disagreement among scientists has made science a powerful force.

Disagreements created by environmental issues, however, are not just among scientists. Dependence on fairly conducted scientific debate is, of course, much preferable to dependence on fear-instilling prophecies. However, due to a possible over-riding and controlling political consensus, scientific consensus often faces a futile scientific debate. The *science über alles* maxim that directs researchers dictates that scientists must continue to strive for the truth that they represent and seek. If satellite monitoring can maintain scientific integrity (preferably backed by definitive science, but possibly supported by a compromise between defendable science and its defendable usage) while disclaiming a popular fervor or supporting an unpopular one, it is duty-bound to be confrontational. Remote sensing and spectro-optical modeling of complex waters did succeed in at least partially closing the

Pandora's box of escaping UV-B demons allegedly attacking inland waters through ever expanding ozone "holes."

However, remote sensing should not become confrontational, if, as will probably most often be the case, no definitive support or lack of support can emerge from analyses of aircraft or satellite data on their own. Most unsettled controversies will involve time-series space data and models that infer biology from physics as well as models that predict how that inferred biology will translate into societal and/or environmental health and behavior. Reiterating: in the environment, almost everything varies in space and time and almost everything is related to almost everything else.

Thus, values of any two *different* parameters may be meaninglessly plotted against one other. Also any two measurements of the *same* parameter taken at two different times may be meaninglessly intercompared. In the time–space continuum of climate variable data (temperature, wind speed, precipitation, ice cover, and geographic size), numerical values may be readily found or inferred to indicate (with little to no statistical reliability) that a variable of choice is increasing, decreasing, or unvarying with time and/or space. Selective data may be easily obtained to support selective belief in selective policy.

Nonetheless, whether or not remote sensing can support or refute an already enshrined policy, its findings and products must be reputable and available to and readily understood by environmental managers and political policy-makers. Enigmatically, salvation of remote sensing relies on adherence to the "science in support of policy" reality. Policy-driven science can be perverted by dictates from the mind-sets of publicly influential nonscientists and these dictates may be detrimental to scientific integrity. It is, therefore, hazardous that, in a time period when sociopolitics and liberal arts are capable of usurping dispassionate science, we are advocating that environmental monitoring from space consider a compromise between defendable science and defendable usage of science. This might appear akin to leading a lamb to slaughter or, at least, to leading a lamb to use as a "selective science" weapon for ecocrusading activists.

Obviously, this is not our intention and extreme caution must be taken to ensure that it does not happen. Remote sensing is continuing to shed the remnants of a handicapping cloak of disrespectability woven from numerous prior incantations. More such incantations would be counterproductive to the shedding process. Although space monitoring of inland and coastal waters is not yet as operational as space monitoring of mid-ocean waters, it is an equal partner. Limnologists, oceanographers, space agencies, governments, universities, and the private sector are engaging in collaborative approaches to space monitoring of water quality. However, the sponsoring public (the actual public, not the self-appointed pop-culture icons, activists, and media hounds that "speak" for it) has yet to be informed, convinced, and brought into the collaboration. Regardless of environmental activists' forewarnings of impending doom, ecosystem monitoring from space can ill afford to don a heavier cloak of disrespectability. Products emerging from the space monitoring of inland and coastal waters must be reliable and trustworthy.

Seductive, truth-bending, targeted consumer advertising may not be detrimental to promotion of throw-away pop-culture fad products. However, from its checkered past of nondefendable fabulist claims and endorsements, remote sensing has already produced its share of spurious throw-away pop-culture fad products. Barnum and Bailey are inappropriate barkers for space-acquired water quality products. To survive, remote sensing must adhere to the "truth in advertising" maxim when promoting the value-added nature of its products. The LRTAP experience of Section 7.3 is a valuable lesson. If space monitoring of inland and coastal water quality is destined for premature consignment to history, let it not be by self-consignment. At a time when climate and environmental changes are emotionally contested issues, it is highly unlikely that remote sensing would recover from a second bout of self-sabotage.

It is urgent that the atmospheric correction dilemma be solved. Nevertheless, remote sensing science and remote sensing technology are sound. Although no product should ever be considered optimal, no legerdemain or magical spells are needed for promoting the value-added nature of space-derived water quality products. Closing on a somewhat comforting note, therefore, remember that the Wizard of Oz did not give Dorothy and her traveling companions on the yellow-brick road anything that they had not already developed and possessed. *Eres tú!*

> This above all: to thine own self be true/And it must follow, as the night the day/Thou canst not then be false to any man.
>
> **From *Hamlet, Prince of Denmark*, Act 1, Scene 3, William Shakespeare, 1603**

Acronyms

ADEOS	Advanced Earth Observing Satellite
AISA	Airborne Imaging Spectrometer
AMRS	Alliance for Marine Remote Sensing
AOP	Apparent Optical Property
ARC	Affiliated Research Center (NASA)
ARIES	Australian Resource Information and Environmental Satellite
ASAR	Advanced Synthetic Aperture Radar
AVHRR	Advanced Very High Resolution Radiometer
AVIRIS	Airborne Visible/Infrared Imaging Spectrometer
BIO	Bedford Institute of Oceanography
BOREAS	Boreal Ecosystem–Atmosphere Study
CASI	Compact Airborne Spectrographic Imager
CCRS	Canada Center for Remote Sensing
CDOM	Colored Dissolved Organic Matter
CESL	Canadian Expert System Laboratory
CPA	Color Producing Agent
CSA	Canadian Space Agency
CSIRO	Commonwealth Scientific and Industrial Research Organization
CWS	Canadian Wildlife Service
CZCS	Coastal Zone Color Scanner
DCM	Deep Chlorophyll Maximum
DEM	Digital Elevation Model
DFO	Fisheries and Oceans Canada
DOC	Dissolved Organic Carbon
DOM	Dissolved Organic Matter
EC	Environment Canada
EEZ	Exclusive Economic Zone (200 nautical miles)
ENSO	El Niño Southern Oscillation
ENVISAT	Environment Satellite
EO	Earth Observation
EOADP	Earth Observation Applications Development Program

EOP	Earth Observation Platform
EOS	Earth Observation System
ERTS	Earth Resources Technology Satellite
ESA	European Space Agency
ESARP	Earth Science Applications Research Program
FLH	Fluorescence Line Height
FLI	Fluorescence Line Imager
GCM	General Circulation Model
GCOS	Global Climate Observing System
GEO	Group on Earth Observations
GEWEX	Global Energy and Water Cycle Experiment
GIS	Geographic Information System
GLERL	Great Lakes Environmental Research Laboratory
GLI	Global Imager
GOOS	Global Ocean Observing Systems
GRIP	Government Related Initiatives Program
GTOS	Global Terrestrial Observing System
HABs	Harmful Algal Blooms
HCFC	Hydrochlorofluorocarbon
HERO	Hyperspectral Environment and Resource Observer
HYMAP	Airborne Hyperspectral Mapper
IGBP	International Geosphere–Biosphere Program
IJC	International Joint Commission
IOCCG	International Ocean Color Coordinating Group
IOP	Inherent Optical Property
IOS	Institute of Ocean Sciences
ISA	Indian Space Agency
ISESS	International Symposium on Environmental Software Systems
ISS	Information Support System
I-STOP	Integrated Satellite Targeting of Polluters
LANDSAT	Land Remote Sensing Satellite
LMA	Lake Management Authority
LOICZ	Land–Ocean Interactions in the Coastal Zone
LRTAP	Long-Range Transport of Air Pollutants
LSPIM	Land Surface Processes and Interaction Mission
LTSP	Long-Term Space Plan (Canadian Space Agency)
MERIS	Medium Resolution Imaging Spectroradiometer
MODIS	Moderate Resolution Imaging Spectroradiometer
MSC	Meteorological Service of Canada
MSS	Multispectral Scanner
NASA	National Aeronautics and Space Administration
NDVI	Normalized Difference Vegetation Index
NEAR	NASA Rendezvous Mission to Near-Earth Asteroid
NEMO	Naval Earth Map Observer
NOAA	National Oceanic and Atmospheric Administration

NRL/SSC	Naval Research Laboratory/Stennis Space Center
NWRI	National Water Research Institute
OCTS	Ocean Color and Temperature Scanner
PAR	Photosynthetic Available Radiation
PCBs	Polychlorinated Biphenyls
POES	Polar Orbiting Environmental Satellite
POLDER	Polarization and Directionality of Earth's Reflectances
QAA	Quantum Adiabatic Algorithm
RADARSAT	Radio Detection and Ranging Satellite (RADAR Satellite)
RESAC	Regional Earth Science Applications Center (NASA)
ROSIS	Reflective Optics System Imaging Spectrometer
SAR	Synthetic Aperture Radar
SeaWiFS	Sea-Viewing Wide Field-of-View Sensor
SIMSA	Spectral Imaging Mission for Science and Application
SMARTSAT	Student Mentored Advanced Research and Technology Satellite (Program)
SOM	Suspended Organic Matter
SPOT	Systeme Probatoire d'Observation de la Terre
TM	Thematic Mapper
TOA	Top of Atmosphere
TOMS	Total Ozone Mapping Spectrometer
TOS	The Oceanographic Society
UN	United Nations
VARS	Value-Added Remote Sensing
WATERS	Water-to-Air Transference of Electromagnetic Radiation System
WCRP	World Climate Research Program
WHYCOS	World Hydrological Cycle Observing System
WMO	World Meteorological Organization
WQMA	Water Quality Management Authority

Glossary

Absorption Coefficient $a(\lambda)$ The fraction of radiant energy absorbed from an incident light beam as it traverses an infinitesimal distance, divided by that infinitesimal distance.

Absorption Cross Section $a_i(\lambda)$ See specific absorption coefficient $a_i(\lambda)$.

Acid Rain (Acid Deposition) A broad and general term that refers to wet or dry acidic substances that fall from the atmosphere. Wet deposition refers to acidic rain, fog, and snow. Dry deposition refers to acidic particles and gases. Principal causes of acid rain are injections, from energy generation processes, of sulphur dioxide (SO_2) and nitrogen oxides (NO_X) that react with atmospheric water, oxygen, and other indigenous chemicals to form acid compounds. Generally, acid rain is a mild solution of sulphuric acid, H_2SO_4, and nitric acid, HNO_3; therefore, it is a serious environmental problem.

Action Spectrum The contribution to a photoprocess of each wavelength of an incident irradiance spectrum. Although action spectra are used to describe photoprocesses that are photochemical or photobiological in nature, the term most commonly refers to an *organism's* spectral response function to an incident irradiance spectrum.

Active Remote Sensing Device A remote sensor that transmits a signal to the target being monitored and then measures the signal that is returned from that target. Active remote sensing devices interact with the target — often stimulating an optical transition within the target resulting in an optical return that would not have been observed from a nonstimulated target. Active remote sensing devices include fluorosensors, scatterometers, and synthetic aperture radar (SAR) systems.

Aerosol Optical Depth An estimate of atmospheric aerosol size and density distributions, defined as the integration of the vertical atmospheric beam attenuation coefficient over altitude h.

Aerosol Phase Function A measure of light scattered by an atmospheric aerosol particle as a function of angle relative to the direction of the incident photon beam, defined as the ratio of the volume scattering function to the atmospheric beam single-scattering coefficient normalized to yield a total scattering probability of 4π.

Aerosols Suspended atmospheric matter and liquid particles that may exist in a myriad of diverse forms and shapes from a myriad of diverse sources. Aerosols include smoke, water and H_2SO_4 droplets, dust, ashes, pollen, spores, and other forms of atmospheric suspensions.

Albedo $A(\lambda)$ The ratio of the energy returning from a point or a surface to the energy incident upon that point or surface for a particular wavelength, λ.

Algae The most familiar of the genus of phytoplankton. Algae are a high-variety group of plant structures that may be one-celled, colonial, or filamentous; they include such diverse members as seaweed and pond scum.

Algal Bloom Fast growing aquatic phytoplankton or algal species that accumulate into dense surficial concentrated patches that generally impart specific colors to the water as viewed from above. Algal blooms may be toxic (harmful algal blooms, or HABs), benign, beneficial, or generally of nuisance. Inland and near-coastal water algal blooms are very often blue-green algal blooms and are invariably toxic. Ocean water algal blooms are very often red tides and may be toxic or nontoxic.

Analytic Model or Algorithm A conceptual or mathematical relationship formulated from well-established scientific principles.

Apparent Optical Property An optical property of a water body that depends upon the spatial distribution of the incident radiation. Apparent optical properties include the irradiance attenuation coefficient and the volume reflectance.

Archimedes' Spiral An ever expanding spiral of polar radius, r, defined by the linear mathematical polar equation $r = k\theta$, where θ is the polar angle cycling through the range 0 to 360°. The effluent of a circular rotating garden hose when viewed from directly overhead is a classic illustration of an Archimedes' spiral. Viewing the solar magnetic field lines frozen within the solar plasma wind from directly overhead and perpendicular to the plane of the ecliptic is a more dramatic illustration of an Archimedes' spiral. Such a configuration is a geophysical feature often visible in the Point Pelee region of Lake Erie due to interplay of directions, speeds, and loadings of Lake Erie lake currents.

Atmospheric Attenuation Coefficient The fraction of radiant energy removed from a solar beam as it traverses an infinitesimal distance (due to the combined processes of scattering and absorption within the atmosphere), divided by that infinitesimal distance. It is the atmospheric counterpart of attenuation coefficient $c(\lambda)$.

Attenuation Coefficient $c(\lambda)$ See total (or beam) attenuation coefficient $c(\lambda)$.

Backscattering Coefficient $b_B(\lambda)$ The integral of the volume scattering function $\beta(\theta, \Phi)$ over the hemisphere trailing the incident flux [defined by the angular ranges $(\pi/2 \leq \theta \leq \pi)$ and $(0 \leq \Phi \leq 2\pi)$]. The product of $B(\lambda)$ and $b(\lambda)$ is b_B.

Backscattering Cross Section $Bi(\lambda)bi(\lambda)$ or $(b_B)i$ See specific backscattering coefficient $B_i(\lambda)b_i(\lambda)$ or $(b_B)_i(\lambda)$.

Backscattering Probability $B(\lambda)$ The ratio of the scattering into the hemisphere trailing the incident flux to the total scattering into all directions. $B(\lambda)$ is determined as the ratio of the backscattering coefficient $b_B(\lambda)$ to the scattering coefficient $b(\lambda)$.

Binning A term referring to the handling and storing of information acquired from data sets from disparate sensors of disparate characteristics and disparate analyses schemes (models, mathematical manipulations, data rejection criteria, etc.) by categorizing the data into respective "bins" that will allow utilizing the collected data as an intercomparable long-term data archive.

Bio–geo-optical Models See biophysical models.

Biophysical Models Models that relate the optical return from a natural water body to the biological aquatic components and processes responsible for that optical return. The physical aspect of the models has historically been considered to be optical. However, recognizing the impact of basin geology (suspended organic matter) on inland water color, the more appropriate term would be bio–geo-optical models.

Bioproductivity The rate at which energy is stored within an ecosystem or part thereof during a specified period of time.

Bioproduction The total chemical energy stored within an ecosystem or part thereof at a specified instant of time.

Bragg-Scale Waves Centimeter-scale sea surface roughness waves generated by near-surface wind fields. Synthetic aperture radar (SAR) sensors are very sensitive to backscatter from Bragg-scale waves.

Bulk Inherent Optical Property An inherent optical property of a water column wherein the water column is considered as a composite entity with no regard as to the specific components contributing to that property. Examples of bulk inherent optical properties are $a(\lambda)$, $b(\lambda)$, $c(\lambda)$, and $b_B(\lambda)$.

Case 1 and Case 2 Waters A simplistic bipartite classification of the optical complexities of natural waters. Case 1 waters (archetypical of mid-oceans) are generally optically binary waters whose color is determined by molecular water and indigenous chlorophyll. Case 2 waters (archetypical of most, but not all, inland and coastal waters) are those whose color-producing agents (CPAs) also include land-derived organic and inorganic matter.

Chromaticity The apportionment of white light into its tristimulus red, green, and blue components.

Clarity A rather loosely defined term describing aquatic composition. It is generally considered to refer to the totality of suspended organic and inorganic matter present in the water column, but not aquatic matter in solution. Clarity implies a high degree of optical transmission brought about by very small concentrations of suspended aquatic matter. Clarity is the antonym, or at least the lower limit, of the also loosely defined term, turbidity.

Color-Producing Agents (CPAs) Indigenous organic and inorganic aquatic components whose combined scattering and absorption of incident light is responsible for aquatic color.

Cryosphere The portion of the Earth's surface where water is in its solid form. The cryosphere includes marine and freshwater ice, glaciers, snow, and permafrost.

Detritus Small particles of organic and partially mineralized matter formed from decayed plants and animals and their excretions.

Diffuse Attenuation Coefficient for Downwelling Irradiance $K_d(z,\lambda)$ The rate of decrease in downwelling spectral irradiance with increasing aquatic depth, z.

Diffuse Attenuation Coefficient for Upwelling Irradiance $K_u(z, \lambda)$ The rate of decrease in upwelling spectral irradiance with increasing aquatic depth, z.

Diffuse Light $E_{sky}(\lambda)$ The spectral distribution of sky irradiance. Also referred to as diffuse sky radiation and skylight.

Diffuse Sky Radiation $E_{sky}(\lambda)$ See diffuse light.

Direct (Collimated) Light $E_{sun}(\lambda)$ The spectral distribution of solar irradiance. Also referred to as direct solar radiation.

Direct Solar Radiation $E_{sun}(\lambda)$ See direct (collimated) radiation.

Discharge Area See principal groundwater regimes.

Downwelling Irradiance $E_d(\lambda)$ The irradiance at a point in an attenuating medium due to the stream of downwelling light.

EcoMAP A generic term, used in a variety of acronym guises, by environmental researchers, government agencies, and the private sector to define the biodiversity of a country or region. Continental and coastal ecozones are demarcated by means of compiling a national hierarchy of ecological units that systematically divide the country or region into progressively smaller and smaller areas of similar physical (climate, geology, soils, hydrology) and biological (natural communities of animals, plants, microorganisms) characteristics. Synergistic interplay of social, cultural, economic, and environmental factors is occasionally, but not always, considered in commercially available GIS terrain mapping techniques.

Ecosystem An all-encompassing term that refers to a geographical area as the definitive and synergistic sum of all the living organisms (plants, animals, microbes, people), their surroundings (water, air, soil), and the natural cycles (water, carbon, climate, solar) that sustain them. An aquatic ecosystem can be as small as a single representative of a single species to as large as a major watershed.

Ecotone The transition area between adjacent ecozones (e.g., between forest and wetland; between wetland and standing water; between tundra and forest) that displays characteristics of each ecozone (including plant and animal species).

Ecozone An ecologically distinct area of the Earth's biosphere that has assumed permanently recognizable features due to synergic interplay among geology, water, vegetation, climate, landform, soil, wildlife, and human factors. Examples of ecozones include forest, tundra, wetland, mountain, grassland, and desert.

Elastic Scattering Spectral scattering processes in which scattered photons undergoing attenuation of energy and changes in direction of propagation do not undergo changes in wavelength or polarization. Rayleigh scattering is an elastic process.

El Niño Southern Oscillation (ENSO) A recurrent disruption to the ocean–atmosphere system resulting from weakening in the velocities of trade winds in the central and western Pacific Ocean.

Empirical Model or Algorithm A conceptual or mathematical relationship formulated from sole reliance upon observation and practical experience rather than from well-established scientific principles.

Epilimnion The warmer upper layer of a lake that is seasonally thermally stratified.

Euphotic Depth See photic depth.

Euphotic Zone See photic zone.

Eutrophic See trophic status.

Eutrophication The enrichment of waters with plant nutrients, generally phosphorus and/or nitrogen, resulting in enhanced aquatic plant growth. See trophic status.

Extraterrestrial Radiation $E_{solar}(\lambda)$ The total electromagnetic energy per unit time per unit area normal to the Sun's rays at one astronomical unit and outside the Earth's atmosphere. It is recorded as the total electromagnetic irradiance impinging at the top of the Earth's atmosphere. Within a seasonal variation of ~3.5% resulting from the elliptical orbit of the Earth, $E_{solar}(\lambda)$ may be considered a constant at 1367 W m^{-2}.

Fluorescence A substance's release of electromagnetic radiation of a particular wavelength as a consequence of that substance absorbing electromagnetic radiation of a different wavelength. The stimulated release of electromagnetic radiation ceases immediately upon cessation of input energy. Aquatic chlorophyll and aquatic dissolved organic matter are fluorescing substances that undergo trans-spectral processes.

Fluorescence Quantum Efficiency See fluorescence quantum yield.

Fluorescence Quantum Yield The ratio of the number of fluorescing photons of emission wavelength λ_{em} to the number of absorbing photons of wavelength λ_{ex}.

Forbush Decrease Decrease in the solar and galactic cosmic ray flux arriving at the Earth's atmosphere, generally occurring about a day or two subsequent to a solar flare. This is a consequence of an electromagnetic barrier created by the interaction of solar plasma with the

Earth's magnetic field. Named after its first observer in 1954, Scott Forbush.

Forward Optical Model A bio–geo-optical model that utilizes inherent optical properties and concentrations of color-producing agents to generate water color.

Forward-scattering Coefficient $b_F(\lambda)$ The integral of the volume scattering function $\beta(\theta,\Phi)$ over the hemisphere preceding the incident flux [defined by the angular ranges $(0 \leq \theta \leq \pi/2)$ and $(0 \leq \Phi \leq 2\pi)$].

Forward-scattering Probability $F(\lambda)$ The ratio of the scattering into the hemisphere preceding the incident flux to the total scattering into all directions. $F(\lambda)$ is determined as the ratio of the forward-scattering coefficient $b_F(\lambda)$ to the scattering coefficient $b(\lambda)$.

Geographic Information System (GIS) A system of hardware and software that provides the ability and the facility to input, store, analyze, retrieve, and output spatial information referenced to a geographical location on the Earth's surface.

Global Radiation $E(\lambda)$ The total irradiance reaching a point on the Earth's surface. It is the sum of the direct solar radiation $E_{sun}(\lambda)$ and the diffuse sky radiation $E_{sky}(\lambda)$.

Harmful Algal Blooms (HABs) Dense, visible patches of algal species that produce potent neurotoxins that can invade and be transferred through the aquatic food chain, thereby exerting a negative impact on not only the lower end of the aquatic food web, but also higher forms of life such as shellfish, sea-birds, sea-mammals, humans, and wildlife. Invariably, algal blooms associated with anthropogenic run-off into lakes, rivers, streams, and coastal regions are HABs.

Humic Acid A water-soluble fraction of humic substances present in the water column as dissolved organic carbon.

Humic Substances A variety of complex polymers (comprising a water-soluble and a water-insoluble fraction) resulting from the decomposition of phytoplankyon cells. The water-soluble fraction gives rise to yellow substance, thus imparting a yellow hue to natural waters.

Humolimnic Acid See yellow substance.

Hydrologic Cycle The circulation and conservation of the Earth's natural water supply. The cycle begins with evaporation of water from the ocean surface. Through a repetitive series of condensation, transport, evaporation, precipitation, infiltration, and transpiration processes around the globe (involving standing water, the atmosphere, and the ground), this cycle returns water to the ocean. The hydrologic cycle is commonly referred to as the water cycle.

Hypertrophic See trophic status.

Hypolimnion The cooler bottom layer of a lake that is seasonally thermally stratified.

Inelastic Scattering Spectral scattering processes wherein scattered photons, in addition to undergoing attenuation of energy and changes in direction of propagation, also undergo changes in wavelength and polarization. Raman scattering is inelastic and, therefore, a trans-spectral process.

Inherent Optical Property (IOP) An optical property of a water body that is totally independent of the spatial distribution of the impinging radiation. Inherent optical properties include the attenuation coefficient, absorption coefficient, scattering coefficient, forward-scattering probability, backscattering probability, and scattering albedo.

Inverse Optical Model A bio–geo-optical model that utilizes water color and IOPs of indigenous CPAs to generate the coextant concentrations of the CPAs *or* water color and the coextant concentrations of CPAs to generate the IOPs of those CPAs.

Irradiance $E(\lambda)$ The radiant flux per unit area at a point within a radiative field or at a point on an extended surface.

Irradiance Attenuation Coefficient $k(\lambda,z)$ The logarithmic depth derivative of the spectral irradiance $E(\lambda,z)$ at subsurface depth, z. See attenuation coefficient.

Irradiance Reflectance $R(\lambda,z)$ The ratio of the upwelling irradiance at a point in an attenuating medium to the downwelling irradiance at that point. In water, this ratio is termed subsurface reflectance or volume reflectance.

Jet Stream A high-altitude, discontinuous band of fast flowing air that brings about significant changes in local weather on scales of hours to days. With fluctuating north–south oscillations, jet streams generally move from west to east over northern mid-latitudes.

Junk Science A pejorative, controversial, and confrontational term indiscriminately and highly judgmentally used to discredit science that is used as evidence (correctly, circumstantially, or fraudulently) to advance a result or conclusion in conflict with the belief of the term's user. As used in this book (admittedly pejorative and controversially judgmental), it is regarded as unfounded opinion through which a sincere, albeit gullible, audience is intentionally betrayed by purported spin rather than scientific substance, although it has sometimes been ameliorated as "judgmentally selective science." Junk science is not to be confused with the legitimacy of the term "empirical," which refers to conceptual inference formulated from sole reliance upon observation and practical experience rather than from well-established scientific principles.

Light A general term, usually referring to radiation in that portion of the electromagnetic spectrum to which the human eye is sensitive (about 390 nm to about 740 nm). See visible wavelengths.

Mesotrophic See trophic status.

Mie Scattering The relationship between scattering intensity and wavelength for the situation in which the wavelength of the impinging radiation is comparable to or smaller than the diameter of the scattering center. That is, Mie scattering involves large particles and thus describes atmospheric aerosol scattering phenomena. Aerosols scatter all visible wavelengths almost equally, thus explaining why clouds appear white.

Monte Carlo Simulation A semiempirical approach to solving mathematically complex equations. Boundary-controlled simulations of large numbers of random occurrences of the consequences of the mathematical formulism are statistically compiled, thereby offering insight into the scientific process defined by the complex equation. An approach long used in theoretical physics (Brownian motion of atoms and molecules; interaction of solar and galactic cosmic radiation with interplanetary magnetic field structures), its widespread use in aquatic optics research has been to simulate photon propagation into, within, and emergent from an attenuating medium. Thus, the complex radiative transfer equation can relate the properties of the radiation field to the inherent absorption and scattering properties of the attenuating medium.

Morphology The science or study that deals with form and structure as related to the entirety of a targeted domain in biology (plants, animals, organisms), geography (landforms, soils), or linguistics (words, syntax).

Normalized Difference Vegetation Index NDVI A simplistic ratio involving near-infrared radiances and visible radiances used to estimate temporal and/or spatial changes in relative terrestrial bioproductivity from satellite altitudes. NDVI is based on near-infrared wavelength reflectance from and visible wavelength absorption by chlorophyll-laden verdure.

Oligotrophic See trophic status.

Optical Cross Section Spectra The specific amount of absorption and scattering, as a function of wavelength λ, that may be attributed to a unit concentration of each organic and inorganic component (CPA) of a natural water body. The absorption, scattering, and backscattering cross sections ($a_i(\lambda)$, $b_i(\lambda)$, and $B_i(\lambda)b_i(\lambda)$) are interchangeable with the terms specific absorption, specific scattering, and specific backscattering coefficients, respectively.

Optical Data In broadest terms, optics is defined as the branch of physics that concerns the interactions of electromagnetic radiation with its containment medium as radiation propagates through that medium. Electromagnetic radiation encompasses a wide spectral range of wavelengths extending from short wavelength gamma radiation ($\lambda < 10^{-5}$ μm) through x-ray, ultraviolet, visible, and infrared radiation to long wavelength microwave radiation ($\lambda > 3000$ μm). Somewhat illogically, common usage has come to refer to a restricted subsection of the electromagnetic spectrum as the optical range (inclusive of ultraviolet, visible, and near-infrared radiation) and to radiance or irradiance collected within the optical range as optical data.

Optical Range A collective term referring to the region of the electromagnetic spectrum that includes the ultraviolet, visible, and near-infrared wavelengths ($0.01\ \mu m < \lambda < 850\ \mu m$).

Optical Transmission The ratio of the light emerging from an infinitesimal attenuating volume to the incident light impinging upon that infinitesimal volume.

Passive Remote Sensing Device A remote sensor that faithfully responds to the optical field emanating from a target whether that optical field be reflected or self-generated. The most familiar passive remote sensing device is the multispectral radiometer.

Penetration Depth $z_p(\lambda)$ The vertical aquatic depth from which 90% of the signal recorded by a remote sensing device is considered to originate. A subsurface irradiance level of ~37% corresponds to $z_p(\lambda)$.

pH A contracted term for p(otential of) H(ydrogen). In chemistry, pH is defined as the negative logarithm (base 10) of the hydrogen–ion concentration, in moles per liter, and is a numerical measure (range $0 \le pH \le 14$) of the relative acidity (pH < 7) or alkalinity (pH > 7) of a solution. A pH of 7 represents neutrality.

Phase Function $P(\theta,\Phi)$ A form of the volume scattering function $\beta(\theta,\Phi)$, normalized to the scattering coefficient $b(\lambda)$, and given as 4π times $\beta(\theta,\Phi)$ divided by $b(\lambda)$. $P(\theta,\Phi)$ is zero for isotropic scattering and is independent of Φ for trans-spectral processes.

Photic Depth The vertical distance from the air–water interface to the 1% subsurface irradiance level. Also called the euphotic depth.

Photic Zone The aquatic region in which maximum photosynthesis occurs (generally taken to be the upper aquatic layer bounded by the 100 and 1% subsurface irradiance levels). Also called the euphotic zone.

Photon A quantum of energy displaying a particulate and a wavelike character. The energy of a photon is given by the product of the speed of light in a vacuum, c (3.00×10^8 m s^{-1}); Planck's constant, h (6.625×10^{-34} J s); and the inverse of its associated wavelength λ.

Photosynthesis The biological combination of chemical compounds in the presence of light. It generally refers to the production of organic substances from carbon dioxide and water residing in green plant cells if the cells are sufficiently irradiated to allow chlorophyllous pigments to convert radiant energy into the chemical compounds.

Photosynthetic Available Radiation (PAR) The total radiation in the wavelength interval of 400 to 700 nm. It is the wavelength interval of greatest significance to aquatic photosynthesis and primary productivity.

Photosynthetic Usable Radiation (PUR) The amount of the photosynthetic available radiation (PAR) that is pertinent to the local photosynthetic process. PUR is the value of PAR as weighted by the absorption capabilities of indigenous aquatic algal species.

Phytoplankton The plant organism component of plankton. See algae.

Phytoplankton Biomass The organic carbon content of the phytoplankton population.

Plankton A collective term representing the principal living organisms (plant, animal, microbial) present within a natural water column. Plankton include phytoplankton (plant), zooplankton (animal), and bacterioplankton (microbial).

Polarization The orientation of the electric field vector in the plane perpendicular to the direction of electromagnetic propagation. Depending upon this orientation, the traveling wave may be plane, circularly, elliptically, or randomly (non) polarized.

Primary Production The chemical energy contained within an ecosystem as a direct result of photosynthesis.

Primary Productivity The sum of all photosynthetic rates within an ecosystem.

Principal Groundwater Regimes The demarcation of a basin or watershed in terms of the depth of the regional water table at a given location within that basin or watershed. The regions of the basin or watershed in which the water table is in close proximity to or at the terrain surface are termed discharge areas. Regions in which the water table is substantially below the terrain surface are termed recharge areas. The regions in which the water table is at intermediate depths and under conditions of excessive precipitation or drought may assume the characteristics of discharge or recharge areas are termed transition areas.

Q (factor) The ratio of the upwelling irradiance just beneath the air–water interface, $E_u(0^-)$, to the upwelling nadir radiance just beneath the air–water interface, $L_u(0^-)$. For most inland and coastal waters, Q can vary in value from ~2.4 to ~5.6 as the solar zenith angle increases from 0 to 80°.

Quantum Adiabatic Algorithm QAA Very briefly, a quantum mechanical approach to optimizing the solution to an inherently difficult problem. It is based on the well-known quantum physics formula in which a Hamiltonian operator acting upon a mathematically defined wave-function yields a field of mathematical entities termed *eigenvalues*. Performing a gradually evolving number of such operations (thus, the "adiabatic" term of QAA) eventually yields an optimal eigenvalue field that best serves the boundary conditions of the problem.

Quantum Efficiency (Fluorescence) The number of quanta (photons) emitted per quantum absorbed in a photoreaction.

Quantum Quanta See photon.

Quantum Yield See quantum efficiency.

Quiescent Sun The steady-state dynamical equilibrium of solar processes that enable solar behavior to be described as a statistically predictable series of recurrent physical phenomena.

Radiance Attenuation Coefficient $K(\lambda, z, \theta, \Phi)$ The logarithmic depth derivative of the radiance at subsurface depth, z. See attenuation coefficient.

Radiance $L(\theta, \Phi, \lambda)$ The radiant flux per unit solid angle, $d\Omega$ (the solid angle lying along a specified direction), per unit surface area, dA (the surface area lying at right angles to the direction of photon propagation), at any point in a radiative field.

Radiant Energy The quantity of energy in joules transferred by radiation. It represents the total energy contained within a beam of photons.

Radiant Flux The time rate of flow of radiant energy.

Radiant Intensity The radiant flux per unit solid angle in a particular direction.

Radiative Transfer The dynamics of the energy changes associated with the propagation of radiation through media that absorb, scatter, and/or emit photons. The radiative transfer equation is the mathematical formulism describing the transfer of energy between an electromagnetic field and a physical containment medium.

Raman Scattering An inelastic scattering process in which an incident light energy excites molecular vibrational modes within the target substance, generally causing the scattered photons to become diminished to energy levels below that of elastic scattering. The vibrational modes are a function of the impinging wavelength and polarization of the target molecule. For visible wavelengths, molecular water scattering phenomena are inelastic and Raman.

Rayleigh Scattering The elastic relationship between scattering intensity and wavelength for the situation in which the wavelength of the impinging radiation is considerably larger than the diameter of the scattering center. That is, Rayleigh scattering involves small particles and thus describes scattering phenomena of atmospheric molecules (oxygen, nitrogen). This scattering is wavelength selective (varying inversely with the fourth power of λ and highly pronounced at the short wavelength region of the visible spectrum, thus explaining why a cloudless sky appears blue).

Recharge Area See principal groundwater regimes.

Red Tides A naturally occurring oceanic algal bloom comprising fast growing red algal species that, when accumulating into dense surficial patches, often impart a reddish color to the water as viewed from above. Red tide is, however, a misnomer because colors other than red may be imparted to the water and there is no relationship between red tides and oceanic tidal motion. Red tides may or may not be classified as harmful algal blooms (HABs).

Reflectivity $R(\lambda)$ The ratio of the radiation returning from a surface to the radiation impinging on that surface for a given wavelength.

Remote Sensing Reflectance $R_{RS}(\theta_A)$ The ratio of the upwelling radiance from a given nadir direction, θ_A, to the downwelling irradiance, both measured at the remote platform.

Scattering Albedo ω_o The number of scattering interactions that occur within a fixed volume of an attenuating medium expressed as a fraction of the total number of interactions (both scattering and absorption) that occur within that fixed volume.

Scattering Coefficient $b(\lambda)$ The fraction of radiant energy scattered from a light beam as it traverses an infinitesimal distance divided by that infinitesimal distance.

Scattering Cross Section $b_i(\lambda)$ See specific scattering coefficient $b_i(\lambda)$.

Scattering Phase Function $P(\theta,\varphi)$ See phase function.

Secchi Depth S The depth at which a lowered disk vanishes from view of an observer situated above the air–water interface.

Selective Science As used here, a euphemism. See junk science.

Semianalytic Model or Algorithm A conceptual or mathematical relationship formulated by some combination of well-established scientific principles, direct observation, and practical experience. Same as semiempirical model or algorithm.

Semiempirical Model or Algorithm See semianalytic model or algorithm.

Sheen See slick.

Skylight See diffuse light.

Slick On an aquatic surface, a slick is an extended region that displays an essentially constant reflected radiation return over an extended range of wavelengths. For example, in the visible radiation spectra, the presence of oil on water displays a slick that is uniformly less dark than the water upon which it is buoyed. At radar frequencies, the presence of oil on water displays a slick that is darker than the sustaining water. A slick is generally considered an interchangeable term with sheen.

Solar Activity The totality of relatively minor disturbances on the solar surface produced by sunspots (relatively cold regions in the solar photosphere that are transient and nonstationary, but statistically predictable, phenomena). Sunspots do not have a dramatic impact on the electromagnetic solar spectrum observed on Earth. They do, however, have a profound impact upon the corpuscular radiation arriving at the Earth as well as upon the magnetic field configurations comprising the inner solar cavity — consequences of short-lived phenomena known as solar flares. Sunspots follow a roughly 28-day repetition cycle as observed from Earth and are directly responsible for the 11-year solar activity cycle. See quiescent sun.

Solar Zenith Angle θ As measured positively from the vertical, the angle that locates the position of the sun. The zenith angle is a function of local time.

Specific Absorption Coefficient $a_i(\lambda)$ The absorption at wavelength λ that may be attributed to a unit concentration of aquatic component i. Also referred to as the absorption cross section of component i.

Specific Backscattering Coefficient $B_i(\lambda)b_i(\lambda)$ **or** $(b_B)_i(\lambda)$ The backscattering at wavelength λ that may be attributed to a unit concentration of aquatic component i. Also referred to as the backscattering cross section of component i.

Specific Inherent Optical Properties (IOPs) The inherent spectro-optical properties of a water column that can be attributed to the individual scattering and absorption color-producing agents comprising the water column. The optical cross section spectra are specific inherent optical properties.

Specific Scattering Coefficient $b_i(\lambda)$ The scattering at wavelength λ that may be attributed to a unit concentration of aquatic component i. Also referred to as the scattering cross section of component i.

Stratosphere The layer of the Earth's atmosphere (roughly between heights of 15 and 50 km) in which atmospheric temperature increases with increasing height. The stratosphere contains the maximum concentration of atmospheric ozone.

Subsurface Irradiance Reflectance $R(\lambda,z)$ See volume reflectance.

Suspended Organic Matter SOM Nonliving organic matter suspended in natural water bodies. Detritus is the most obvious example of SOM.

Thermal Bar The demarcation between the elevated nearshore aquatic temperature and the as yet to be elevated offshore colder temperature as seasonal thermal lake stratification develops.

Thermocline The region of demarcation between the distinctively warmer upper thermal regime (epilimnion) and the distinctively cooler lower thermal regime (hypolimnion) comprising a thermally stratified lake.

Total (or Beam) Attenuation Coefficient $c(\lambda)$ The fraction of radiant energy removed from an incident light beam as it traverses an infinitesimal distance (due to the combined processes of absorption and scattering) divided by that infinitesimal distance. Also referred to as attenuation coefficient $c(\lambda)$.

Total Scattering Coefficient $b(\lambda)$ See scattering coefficient.

Transition Area See principal groundwater regimes.

Transmittance The ratio of the radiant flux within a beam emerging from an infinitesimal attenuating layer to the incident radiant flux within the beam impinging upon that infinitesimal layer.

Trans-spectral Processes Optical processes in which stimulation by energy at one wavelength (λ_1) leads to release of energy at another wavelength (λ_2), where, in general, $\lambda_2 > \lambda_1$, i.e., a decrease in energy. Raman scattering and fluorescence are trans-spectral processes.

Trophic Levels The succession of living plant and animal species that constitute the life cycle dynamics (or food chain) of a natural water column. Trophic levels represent the nutrient–life–death–decomposition relationships of the water column.

Trophic Status A qualitative term defining the nutrient and plant growth rate conditions of natural water bodies. A water body displaying enhanced nutrient concentrations and plant growth is termed eutrophic; a water body displaying low nutrient concentrations and plant growth is termed oligotrophic; and a water body displaying intermediate nutrient concentrations and plant growth is termed mesotrophic. Hypertrophic and ultraoligotrophic refer to extended boundary conditions of eutrophic and oligotrophic, respectively.

Tsunami (Japanese for "harbor wave") A train of waves in a water body that is not generated by wind stress, but rather by vertical displacements within the water resulting from such disturbances as submarine earthquakes, landslides, volcanic eruptions, explosions, or meteorite impacts. Often improperly referred to as "tidal waves" and/or "seismic waves," tsunamis can savagely attack and devastate coastlines.

Turbidity A rather loosely defined term describing aquatic composition. It is generally considered to refer to the totality of suspended organic and inorganic matter present in the water column. It does not include matter in solution. Turbidity implies a low degree of optical transmission brought about by substantial concentrations of suspended aquatic matter. Turbidity is the antonym of the also loosely defined term "clarity."

Ultraoligotrophic See trophic status.

Upwelling Irradiance $E_u(\lambda)$ The irradiance at a point in an attenuating medium due to the stream of upwelling light.

UV Photon Budget As used in this book, the final dispersal or apportionment of subsurface solar UV-A and UV-B photons that have penetrated the atmosphere as well as the air–water interface. The apportionment, as a function of wavelength (290 to 400 nm) is the consequence of absorption by indigenous columnar concentrations of chlorophyll and dissolved organic matter.

Vertical Irradiance Attenuation Coefficient $K(\lambda,z)$ The value of the Irradiance attenuation coefficient when the vertical subsurface depth, z, is considered as opposed to the actual aquatic distance that would be traversed by a photon flux propagating in a nonvertical direction.

Visible Wavelengths The region of the electromagnetic spectrum to which the human eye is considered to be responsive. It is generally taken to encompass the wavelength interval from about 390 to about 740 nm.

Volume Reflectance $R(\lambda,z)$ The ratio of the upwelling irradiance E_u at a point in the water column to the downwelling irradiance E_d at that point. See irradiance reflectance.

Volume Scattering Function $\beta(\theta,\Phi)$ The scattered radiant intensity, dI, in a direction (θ,Φ) per unit scattering volume, dV, normalized to the value of the incident radiation, E_{inc}.

Water-Leaving Radiance $L_u(0^+,\lambda)$ The spectral values of the visible radiance, emergent from the air–water interface, that collectively define the color of the water as seen from immediately above the air–water interface.

Watershed The total land area from which water drains into a single stream, lake, or ocean.

Whitings Extended regions of near-surface milky-white calcium carbonate precipitation resulting from supersaturated solution conditions in natural waters.

Yellow Substance Dissolved aquatic humus that, due to the presence of yellow and brown melanoids (the products of the temperature-dependent Maillard reaction), imparts a yellow hue to the aquatic column. Also referred to as gelbstoff, aquatic humic matter, yellow organic acids, gilvin, and humolimnic acid.

Zooplankton The animal organism component of plankton.

References

Aas, E., Two-stream irradiance model for deep waters, *Appl. Opt.*, 26, 2095–2101, 1987.

Anstee, J., Dekker, A., Byrne, G., Daniel, P., Held, A., and Miller, J., Use of hyperspectral imaging for benthic species mapping in South Australian coastal waters, *Proc. 10th Australasian Remote Sensing Photogrammetry Conf.*, Adelaide, SA, 1051–1061, 2000.

Austin, R.W., The remote sensing of spectral radiance from below the ocean surface, in *Optical Aspects of Oceanography* (Jerlov, N.G. and Steeman–Neilsen, E., Eds.), Academic Press, London, 317–344, 1974.

Babin, M., Morel. A., and Gentili, B., Remote sensing of sea surface Sun-induced chlorophyll fluorescence: consequences of natural variations in the optical characteristics of phytoplankton and the quantum yield of chlorophyll *a* fluorescence, *Int. J. Remote Sens.*, 17, 2417–2448, 1996.

Babin, M. and Stramski, D., Variations in the mass-specific absorption coefficient of mineral particles suspended in water, *Limnol. Oceanogr.*, 49, 756–767, 2004.

Bannister, T.T., Production equations in terms of chlorophyll concentration, quantum yield, and upper limit to production, *Limnol. Oceanogr.*, 19, 1–12, 1974.

Bannister, T.T. and Laws, E.A., Modelling phytoplankton carbon metabolism, in *Primary Productivity in the Sea* (Falkowski, P.G., Ed.), Plenum Press, New York, 243–256, 1980.

Bannister, T.T., Empirical equations relating scalar irradiance to a, b/a, and solar zenith angle, *Limnol. Oceanogr.*, 34, 173–177, 1990.

Barber, D.G., Hochheim, K.P., Dixon, R., Mosscrop, D.R., and McMullan, M.J., The role of Earth observation technologies in flood mapping: a Manitoba case study, *Can. J. Remote Sens.*, 22, 137–143, 1996.

Barrett, E.C., Beaumont, M.J., and Herschy, R.W., Satellite remote sensing for operational hydrology: present needs and future opportunities, *Remote Sens. Rev.*, 4, 351–466, 1990.

Bartlett, J.S., Voss, K.J., Sathyendranath, S., and Vodacek, A., Raman scattering by pure water and seawater, *Appl. Opt.*, 37, 3324–3332, 1998.

Bartley, W.C., Bukata, R.P., McCracken, K.G., and Rao, U.R., Anisotropic cosmic radiation fluxes of solar origin, *J. Geophys. Res.*, 71, 3297–3304, 1966.

Bate, R., Science under siege: the decoupling of science from climate policy, *Energy Environ.*, 7, 323–331, 1996.

Beeton, A.M., Relationship between Secchi disk readings and light penetration in Lake Huron, *Trans. Am. Fish Soc.*, 87, 73–79, 1957.

Berger, W.H., Fischer, K., Lai, C., and Wu, G., Oceanic primary productivity and organic carbon flux. Part 1, overview and maps of primary production and export, *Scripps Inst. Oceanogr.*, (8730), 1–67, 1987.

Binding, C.E., Bowers, D.G., and Michelson–Jacob, E.G., An algorithm for the retrieval of suspended sediment concentrations in the Irish Sea from SeaWiFS ocean color satellite imagery, *Int. J. Remote Sens.*, 24, 3791–3806, 2003.

Birkeland, C. (Ed.), *Life and Death of Coral Reefs*, Chapman & Hall, London U.K., 536 pp., 1997.

Boehmer–Christiansen, S., Political pressure in the formation of scientific consensus, *Energy Environ.*, 365–376, 1996.

Borstad, G.A., Gower, J.F.R., and Bukata, R.P., *The Red Nymph: a small and simple red imager for the International Space Station (ISS)*, suggestion to the Canadian Space Agency, 5 pp., 2001.

Borstad, G.A., Brown, L., Bukata, R.P., Jerome, J.H., Gower, J.F.R., Charlton, M.N., Booty, W.G., Willis, P., Kerr, R., and Horniak, W., *Water Quality Products for Inland and Coastal Canada: Development of Hyperspectral Applications under the Canadian Space Agency's Earth Observations Development Program (EOADP)*, final report, 72 pp., 2002a.

Borstad, G.A., Brown, L., Gower, J.F.R., Bukata, R.P., and Jerome, J.H., *A New Water Quality Information Product for Coastal and Inland Waters*, poster, ASLO summer meeting, Victoria, B.C., June 10–14, 2002b.

Bowers, D.G., Harker, G.E.L., and Stephan, B., Absorption spectra of inorganic particles in the Irish Sea and their relevance to remote sensing of chlorophyll, *Int. J. Remote Sens.*, 17, 2449–2460, 1996.

Brakenridge, G.R., Tracy, B.T., and Knox, J.C., Orbital SAR remote sensing of a river flood wave, *Int. J. Remote Sens.*, 19, 1439–1445, 1998.

Brando, V.E. and Dekker, A.G., The fluorescence term on the observed 690–710-nm reflectance peak in eutrophic turbid (inland) waters: myth or reality? in *Ocean Optics XVI, Proc. SPIE* (Ackelson, S. and Trees, C., Eds.), Office of Naval Research, 1–12, 2002.

Bricaud, A., Morel, A., and Prieur, L., Absorption by dissolved organic matter of the sea (yellow substance) in the UV and visible domains, *Limnol. Oceanogr.*, 26, 43–53, 1981.

Bricaud, A., Morel, A., and Prieur, L., Optical efficiency factors of some phytoplankters, *Limnol. Oceanogr.*, 28, 816–832, 1983.

Bruton, J.E., Jerome, J.H., and Bukata, R.P., Satellite observations of sediment transport patterns in the Lac Saint–Pierre region of the St. Lawrence River, *Water Poll. Res. J. Can.*, 23, 243–252, 1988.

Bukata, R.P., Rao, U.R., McCracken, K.G., and Keath, E.P., Observation of solar particle fluxes over extended solar longitudes, *Solar Phys.*, 33, 229–240, 1972.

Bukata, R.P., Haras, W.S., and Bruton, J.E., Space observations of lake coastal processes in Lake Huron and Lake St. Clair, *Proc. 2nd Can. Symp. Remote Sens.*, 531–549, 1974.

Bukata, R.P., Haras, W.S., and Bruton, J.E., The application of ERTS-1 digital data to water transport phenomena in the Point Pelee–Rondeau area, *Verh. Int. Verein. Limnol.*, 19, 168–178, 1975.

Bukata, R.P., Bruton, J.E., Jerome, J.H., Bobba, A.G., and Harris, G.P., The application of Landsat-1 digital data to a study of coastal hydrography, *Proc. 3rd Can. Symp. Remote Sens.*, 331–348, 1976.

Bukata, R.P., Bobba, A.G., Bruton, J.E., and Jerome, J.H., The application of apparent radiance data to the determination of groundwater flow pathways from sat ellite altitudes, *Can. J. Spectrosc.*, 23, 79–91, 1978.

Bukata, R.P., Jerome, J.H., Bruton, J.E., and Jain, S.C., Nonzero subsurface irradiance reflectance at 670 nm from Lake Ontario coastal waters, *Appl. Opt.*, 18, 3926–3932, 1979.

Bukata, R.P., Jerome, J.H., Bruton, J.E., Jain, S.C., and Zwick, H.H., Optical water quality model of Lake Ontario. 1. Determination of the optical cross sections of organic and inorganic particulates in Lake Ontario, *Appl. Opt.*, 20, 1696–1703, 1981a.

Bukata, R.P., Jerome, J.H., Bruton, J.E., Jain, S.C., and Zwick, H.H., Optical water quality model of Lake Ontario. 2. Determination of chlorophyll *a* and suspended mineral concentrations of natural waters from submersible and low-altitude optical sensors, *Appl. Opt.*, 20, 1704–1714, 1981b.

Bukata, R.P., Bruton, J.E., and Jerome, J.H., The futility of using remotely determined chlorophyll concentrations to infer acid stress in lakes, *Can. J. Remote Sens.*, 8, 38–41, 1982.

Bukata, R.P., Bruton, J.E., and Jerome, J.H., Use of chromaticity in remote measurements of water quality, *Remote Sens. Environ.*, 13, 161–177, 1983.

Bukata, R.P., Bruton, J.E., and Jerome, J.H., *Application of Direct Measurements of Optical Parameters to the Estimation of Lake Water Quality Indicators*, Environment Canada Inland Waters Directorate Scientific Series No. 40, 35 pp. 1985.

Bukata, R.P., Jerome, J.H., and Bruton, J.E., Particulate concentrations in Lake St. Clair as recorded by a shipborne multispectral optical monitoring system, *Remote Sens. Environ.*, 25, 201–229, 1988a.

Bukata, R.P., Jerome, J.H., and Bruton, J.E., Relationships among Secchi disk depth, beam attenuation coefficient, and irradiation attenuation coefficient for Great Lakes waters, *J. Great Lakes Res.*, 14, 347–355, 1988b.

Bukata, R.P., Bruton, J.E., Jerome, J.H., and Haras, W.S., An evaluation of the impact of persistent water level changes on the areal extent of Georgian Bay/North Channel marshlands, *Environ. Manage.*, 12, 359–368, 1988c.

Bukata, R.P., Jerome, J.H., Kondratyev, K.Ya., and Pozdnyakov, D.V., Estimation of organic and inorganic matter in inland waters: optical cross sections of Lakes Ontario and Ladoga, *J. Great Lakes Res.*, 17, 461–469, 1991a.

Bukata, R.P., Jerome, J.H., Kondratyev, K.Ya., and Pozdnyakov, D.V., Satellite monitoring of optically active components of inland waters: an essential input to regional climate change impact studies, *J. Great Lakes Res.*, 17, 470–478, 1991b.

Bukata, R.P., Bruton, J.E., and Jerome, J.H., Utilizing vegetation vigor as an aid to assessing groundwater flow pathways from space (in Russian), *Stud. Earth Space*, 2, 107–118, 1991c.

Bukata, R.P., Jerome, J.H., Kondratyev, K.Ya., and Pozdnyakov, D.V., *Optical Properties and Remote Sensing of Inland and Coastal Waters*, CRC Press, Boca Raton, FL, 362 pp., 1995.

Bukata, R.P. and Jerome, J.H., A (nonlinear) perspective on ultraviolet radiation and freshwater ecosystems, *Proc. Workshop Atmospheric Ozone*, Ontario Climate Advisory Committee, Downsview, 137–165, 1997.

Bukata, R.P. and Jerome, J.H., Satellite imagery as cost-accessible time-series data: a comment, *Can. J. Remote Sens.*, 24, 206–208, 1998.

Bukata, R.P., Whither remote sensing of inland and coastal waters? guest editorial comment, *Backscatter*, spring/summer issue, 8–9, 2001.

Bukata, R.P., Pozdnyakov, D.V., Jerome, J.H., and Tanis, F.J., Validation of a radiometric color model applicable to optically complex water bodies, *Remote Sens. Environ.*, 77, 165–172, 2001a.

Bukata, R.P., Jerome, J.H., and Dekker, A.G., Remote sensing of inland and coastal waters: 2. What do end-users want, what do they need, and what can they be given? Unpublished presentation at the *Workshop on Remote Sensing & Resource Management in Nearshore and Inland Waters*, Wolfville, Nova Scotia, October 22–24, 2001b.

Bukata, R.P., Dekker, A.G., and Borstad, G.A., A clear view of turbid waters, *Backscatter*, winter issue, 36–38, 2002.

Bukata, R.P., Jerome, J.H., Borstad, G.A., Brown, L.N., and Gower, J.F.R., Developing useful inland water quality products from hyperspectral satellite imagery: some early lessons from MODIS and WATERS data, *Backscatter*, winter issue, 29–31, 2003.

Bukata, R.P., Jerome, J.H., Borstad, G.A., Brown, L.N., and Gower, J.F.R., Mitigating the impact of trans-spectral processes on multivariate retrieval of water quality parameters from case 2 waters, *Can. J. Remote Sens.*, 30, 8–16, 2004.

Bukata, R.P. and Helbig, J.A., Monitoring coastal ocean and inland water quality data from hyperspectral water colour data, *Backscatter*, winter issue, 33–37, 2004.

Butler, A.J. and Jernakoff, P., *Seagrass in Australia: Strategic Review and Development of an R&D Plan*, CSIRO Publishing, 210 pp., 1999.

Calkins, J. (Ed.), *The Role of Solar Ultraviolet Radiation in Marine Ecosystems*, Plenum Press, New York, 724 pp., 1982.

Calkins, J. and Barcelo, J.A., Action spectra, in *The Role of Solar Ultraviolet Radiation in Marine Ecosystems*, (Calkins, J., Ed.), Plenum Press, New York, 143–150, 1982.

Campbell, J.W. and O'Reilly, J.E., Role of satellites in estimating primary productivity on the Northeast Atlantic Continental Shelf, *Cont. Shelf Res.*, 8, 179–204, 1988.

Campbell, J.W., Blaisdell, J.M., and Darzi, M., Level-3 SeaWiFS data products: spatial and temporal binning algorithms, in *SeaWiFS Technical Report Series* (Hooker, S.B., Firestone, E.R., and Acker, J.G., Eds.), NASA Technical Memorandum 104566, 32, 73 pp., 1995.

Carder, K.L., Steward, R.G., Harvey, G.R., and Ortner, P.B., Marine humic and fulvic acids: their effect on remote sensing of ocean chlorophyll, *Limnol. Oceanogr.*, 34, 68–81, 1989.

Carder, K.L., Hawes, S.K., Baker, K.A., Smith, R.C., Steward, R.G., and Mitchell, B.G., Reflectance model for quantifying chlorophyll *a* in the presence of productivity degradation products, *J. Geophys. Res.*, 96, 599–611, 1991.

Charlton, M.N., Did zebra mussels clean up Lake Erie? *Great Lake Res. Rev.*, 5, 11–15, 2001.

Charlton, M.N., L'Italien, S., Howell, T., Bertram, P., Zarull, M., Thoma, R., and Culver, D., *Review of Eutrophication and Undesirable Algae in Lake Erie*, Environment Canada National Water Research Institute, Burlington/Saskatoon, NWRI contribution no. 01-177, 2001.

Chipman, J.W., Lillesand, T.M., Schmaltz, J.E., Leale, J.E., and Nordheim, M.J., Mapping lake water clarity with Landsat images in Wisconsin, U.S.A., *Can. J. Remote Sens.*, 30, 1–7, 2004.

Clarke, G.L., Ewing, G.L., and Lorenzen, C.J., Spectra of back-scattered light from the sea obtained from aircraft as a measure of chlorophyll concentration, *Science*, 167, 1857–1866, 1970.

Clover, C., *The End of the Line: How Over-fishing Is Changing the World and What We Eat*, Ebury Press, London, U.K., 314 pp., 2004.

Coble, P.G. and Brophy, M.M., Investigation of the geochemistry of dissolved organic matter in coastal waters using optical properties, in *Ocean Optics XII, Proc. SPIE Int. Soc. Opt. Eng.*, (Jaffe, J.S., Ed.), 55–61, 1996.

Cosper, E.M., Bricelj, M., and Carpenter, E.J. (Eds.), *Novel Phytoplankton Blooms: Causes and Impacts of Recurrent Brown Tide and Other Unusual Blooms*, Springer–Verlag, New York, 799 pp., 1989.

Cullen, J.J., The deep chlorophyll maximum: comparing vertical profiles of chlorophyll *a, Can. J. Fish. Aquat. Sci.*, 39, 791–803, 1982.

Cullen, J.J. and Lesser, M.P., Inhibition of photosynthesis by ultraviolet radiation as a function of dose and dosage rate: results for a marine diatom, *Mar. Biol.*, 111, 183–190, 1991.

Cullen, J.J., Neale, P.J., and Lesser, M.P., Biological weighting function for the inhibition of phytoplankton photosynthesis by ultraviolet radiation, *Science*, 258, 646–650, 1992.

Culver, M.E. and Perry, M.J., Calculation of solar-induced fluorescence in the surface and subsurface waters, *J. Geophys. Res.*, 102(C5), 10563–10572, 1997.

de Mora, S., Demers, S., and Vernet, M. (Eds.), *The Effects of UV Radiation in the Marine Environment*, Cambridge Environmental Chemistry Series 10, Cambridge University Press, New York, 324 pp., 2000.

Dekker, A.G., Detection of optical water quality parameters for eutrophic waters by high resolution remote sensing, doctoral thesis, University of Amsterdam, The Netherlands, 222 pp., 1993.

Dekker, A.G., Malthus, T.J.M., and Hoogenboom, H.J., The remote sensing of inland water quality, in *Advances in Environmental Remote Sensing* (Danson, F.M. and Plummer, E.S., Eds.), Chichester, U.K., John Wiley & Sons, 123–142, 1995.

Dekker, A.G., Hoogenboom, H.J., Volten, H., Schreurs, R., and De Haan, J.F., Angular scattering functions of algae and silt: an analysis of backscattering to scattering fraction, in *Ocean Optics XIII, Pro. SPIE*, (Ackleson, S.G. and Frouin, R., Eds.), 392–400, 1997.

Dekker, A.G., Bukata, R.P., and Jerome, J.H., Remote sensing of inland and coastal waters: 1. Twenty-eight years of technology push but still waiting for end-user pull, unpublished presentation at the *Workshop on Remote Sensing & Resource Management in Nearshore and Inland Waters*, Wolfville, Nova Scotia, October 22–24, 2001a.

Dekker, A.G., Vos, R.J., and Peters, S.W.M., Comparison of remote sensing data, model results, and *in situ* data for total suspended matter (TSM) in the southern Frisian lakes, *Sci. Total Environ.*, 268, 197–214, 2001b.

Dekker, A.G., Peters, S.W.M., Vos, R.J., and Rijkeboer, M., Remote sensing for inland water quality detection and monitoring: state-of-the-art application in Friesland waters, in *GIS and Remote Sensing Techniques in Land- and Water-Management* (van Dijk, A. and Bos, M.G., Eds.), Kluwer Academic Publishers, Dordrecht, The Netherlands, 17–38, 2001c.

Dekker, A.G. and Bukata, R.P., Remote sensing of inland and coastal waters, in *URSI Review of Radio Science 1999–2002* (Hallikainen, M., Ed.), URSI RRS 1999–2002, The Netherlands, chapter 23, 519–534, 2002.

Dekker, A.G., Brando, V.E., Anstee, J.M., Pinnel, N., Kutser, T., Hoogenboom, H.J., Peters, S.W.M., Pasterkamp, R., Vos, R., Olbert, C., and Malthus, T.J.M., Applications of imaging spectrometry in inland, estuarine, coastal and ocean waters, in *Imaging Spectrometry: Basic Principles and Prospective Applications*, (van der Meer, F., Ed.), Kluwer Academic Publishers, Dordrecht, The Netherlands, 2002a.

Dekker, A.G., Brando, V.E., Pinnel, N., and Held, A., Hyperion: Australian scientists take a preliminary look at this imaging spectrometer's application in coastal and inland waters, *Backscatter*, winter issue, 12–15, 2002b.

Dekker, A.G., Anstee, J.M., and Brando, V.E., *Seagrass Change Assessment Using Satellite Data for Wallis Lake, NSW: a Consultancy Report to the Great Lakes Council and Department of Land and Water Conservation*, CSIRO Land and Water, Canberra, Technical Report 13/03, 64 pp., 2003.

Doerffer, R. and Schiller, H., Determination of case 2 water constituents using radiative transfer simulation and its inversion by neural networks, in *Ocean Optics XIV, Pro. SPIE* (Ackleson, S.G. and Campbell, J., Eds.), Office of Naval Research, Washington, D.C., 1998.

Doerffer, R. and Schiller, H., *Determination of water constituents from water leaving radiance reflectances of the ocean color sensor MOS using inverse modeling*, Algorithm Theoretical Basis Document, ESA-ERIN / GMV No. 11672/95/I–HE, 1999.

Doran, P.T., Priscu, J.C., Lyons, W.B., Walsh, J.E., Fountain, A.G., McKnight, D.M., Moorhead, D.L., Virginia, R.A., Wall, D.H., Clow, G.D., Fritsen, C.H., McKay, C.P., and Parsons, A.N., Antarctic climate cooling and terrestrial ecosystem response, *Nature*, 415, 517–520, 2002.

Duane, D.B., Characteristics of the sediment load in the St. Clair River, *Proc. 10th Conf. Great Lakes Res.*, 115–132, 1967.

Duntley, S.Q. and Preisendorfer, R.W., *The Visibility of Submerged Objects*, Mass. Inst. Tech. Visibility Lab. Final Report M5ori-07864, Cambridge, MA, 1952.

Ekelund, N.G.A., Effects of UV-B radiation on growth and motility of four phytoplankton species, *Physiol. Plant.*, 78, 590–594, 1990.

El-Shaarawi, A.H., Sampling strategy for future data collection, in *Statistical Assessment of the Great Lakes Surveillance Program, 1966–1981*, Environment Canada Scientific Series No. 136, chapter 7, 233–264, 1984.

El-Shaarawi, A.H. and Hunter, J.S., Environmetrics, in *Encyclopedia of Environmetrics* (El-Shaarawi, A.H. and Piegorsch, W.W., Eds.), John Wiley & Sons, Chichester, U.K., 2002.

Environment Canada, *A Primer on Water: Questions and Answers*, 2nd ed., 66 pp., 1991.

Environment Canada, *Environment Canada Nature Research Agenda 1999–2004*, 42 pp., 1999a.

Environment Canada, *Environmental Scan for Building the Nature Research Agenda 1999–2005*, 46 pp., 1999b.

Eppley, R.W. and Peterson, B.J., Particulate organic matter flux and planktonic new production in the deep ocean, *Nature*, 282, 677–680, 1979.

Eppley, R.W., Stewart, E., Abbott, M.R., and Heyman, U., Estimating ocean primary production from satellite chlorophyll: introduction to regional differences and statistics for the Southern California Bight, *J. Plankton Res.*, 7, 57–70, 1985.

Fee, E.J., A numerical model for the estimation of photosynthetic production, integrated over time and depth, in natural waters, *Limnol. Oceanogr.*, 14, 906–911, 1969.

Fee, E.J., *Computer Programs for Calculating* in Situ *Phytoplankton Photosynthesis*, Can. Tech. Rep. Fish. Aquat. Sci. No. 1740, 27 pp., 1990.

Fischer, J., On the information content of multispectral radiance measurements over an ocean, *Int. J. Remote Sens.*, 6, 773–786, 1985.

Fischer, J. and Kronfield, U., Sun-stimulated chlorophyll fluorescence. 1. Influence of oceanic properties, *Int. J. Remote Sens.*, 11, 2125–2147, 1990.

Fraser, R.S., Ferrare, R.A., Kaufman, Y.J., and Mattoo, S., Algorithm for atmospheric corrections of aircraft and satellite imagery, *Int. J. Remote Sens.*, 13, 541–557, 1992.

Fraser, R.S., Mattoo, S., Yeh, E.-N, and McClain, C.R., Algorithm for atmospheric and glint corrections of satellite measurements of ocean pigment, *J. Geophys. Res.*, 102, 17107–17118, 1997.

Gallegos, C.L., Correll, D.L., and Pierce, J.W., Modeling spectral diffuse attenuation, absorption, and scattering coefficients in a turbid estuary, *Limnol. Oceanogr.*, 35, 1486–1502, 1990.

Gallie, E.A. and Murtha, P.A., Specific absorption and backscattering spectra for suspended minerals and chlorophyll *a* in Chilko Lake, British Columbia, *Remote Sens. Environ.*, 39, 103–118, 1992.

Gitelson, A.A., The peak near 700 nm on radiance spectra of algae and water: relationships of its magnitude and position with chlorophyll concentration, *Int. J. Remote Sens.*, 13, 3367–3373, 1992.

Gitelson, A.A., Schalles, J.F., Rundquist, D.C., Schiebe, F.R., and Yacobi, Y.Z., Comparative reflectance properties of algal cultures with manipulated densities, *J. Appl. Phytol.*, 11, 345–354, 1999.

Gordon, H.R., Diffuse reflectance of the ocean: the theory of its augmentation by chlorophyll *a* fluorescence at 685 nm, *Appl. Opt.*, 18, 1161–1166, 1979.

Gordon, H.R. and Brown, O.B., Irradiance reflectivity of a flat ocean as a function of its optical properties, *Appl. Opt.*, 12, 1549–1551, 1973.

Gordon, H.R., Brown, O.B., and Jacobs, M.M., Computed relationships between the inherent and apparent optical properties of a flat, homogeneous ocean, *Appl. Opt.*, 14, 417–427, 1975.

Gordon, H.R. and McCluney, W.R., Estimation of the depth of sunlight penetration in the sea for remote sensing, *Appl. Opt.*, 14, 413–416, 1975.

Gordon, H.R. and Wouters, A.W., Some relationships between Secchi depth and inherent optical properties of natural waters, *Appl. Opt.*, 17, 3341–3343, 1978.

Gordon, H.R. and Clark, D.K., Atmospheric effects in the remote sensing of phytoplankton pigments, *Boundary-Layer Meteorol.*, 18, 299–313, 1980.

Gordon, H.R. and Morel, A., Remote assessment of ocean color for interpretation of satellite visible imagery: a review, in *Lecture Notes on Coastal and Estuarine Studies* (Barber, R.T., Mooers, N.K, Bowman, M.J., and Zeitzchel, B., Eds.), Springer–Verlag, New York, 114 pp., 1983.

Gordon, H.R., Clark, D.K., Brown, J.W., Brown, O.B., Evans, R.H., and Broenkow, W.W., Phytoplankton pigment concentrations in the Middle Atlantic Bight: comparison of ship determinations and CZCS estimates, *Appl. Opt.*, 22, 20–36, 1983.

Gordon, H.R., Smith, R.C., and Zaneveld, J.R.V., Introduction to ocean optics, in *Ocean Optics VII, Proc. SPIE* (Blizzard, M.A., Ed.), 489, 2–41, 1984.

Gordon, H.R. and Wang, M., Surface roughness considerations for atmospheric correction of ocean sensors. 1. The Rayleigh scattering component, *Appl. Opt.*, 31, 4247–4253, 1992.

Gordon, H.R. and Wang, M., Retrieval of water-leaving radiance and aerosol optical thickness over the oceans with SeaWiFS: a preliminary program, *Appl. Opt.*, 33, 443–452, 1994.

Gore, A., *Earth in the Balance: Ecology and the Human Spirit,* Houghton–Mifflin, Boston, 407 pp., 1992.

Gould, Jr., R.W. and Arnone, R.A., Remote sensing estimates of inherent optical properties in a coastal environment, *Remote Sens. Environ.*, 61, 290–301, 1997.

Gower, J.F.R. and Borstad, G.A., Use of the *in vivo* fluorescence line at 685 nm for remote sensing surveys of surface chlorophyll, in *Oceanography from Space,* (Gower, J.F.R., Ed.), Plenum Press, New York, 329–338, 1981.

Gower, J.F.R., *Study of Fluorescence-based Chlorophyll* a *Concentration Algorithms for Case 1 and Case 2 Waters,* ESA and ESTEC, contract 12295/97NL/RE, 1999.

Gower, J.F.R., Doerffer, R., and Borstad, G.A., Interpretation of the 685-nm peak in water-leaving radiance spectra in terms of fluorescence, absorption and scattering, and its observation by MERIS, *Int. J. Remote Sens.*, 20, 1771–1786, 1999.

Gower, J.F.R., SeaWiFS global composite images show significant features of Canadian waters for 1997–2001, *Can. J. Remote Sens.*, 30, 26–35, 2004.

Gower, J.F.R., Brown, L., and Borstad, G.A., Observation of chlorophyll fluorescence in west coast waters of Canada using the MODIS satellite sensor, *Can. J. Remote Sens.*, 30, 17–25, 2004.

Graneli, E., Sundstrom, B., Edlar, L., and Anderson, D.M. (Eds.), *Toxic Marine Phytoplankton: Proceedings of the Fourth International Conference on Toxic Marine Phytoplankton,* Elsevier, New York, 1990.

Grassl, H., Pozdnyakov, D.V., Lyaskovsky, A.V., and Pettersson, L., Numerical modeling of trans-spectral processes in natural waters, *Int. J. Remote Sens.*, 23, 1581–1607, 2002.

Green, A.E.S., Cross, K.R., and Smith, L.A., Improved analytic characterization of ultraviolet skylight, *Photochem. Photobiol.*, 31, 59–65, 1980.

Green, A.E.S. and Schippnick, P.F., UV-B reaching the surface, in *The Role of Solar Ultraviolet Radiation in Marine Ecosystems* (Calkins, J., Ed.), Plenum Press, New York, 5–27, 1982.

Green, S. and Blough, N.V., Optical absorption and fluorescence properties of chromophoric dissolved organic matter in natural waters, *Limnol. Oceanogr.*, 39, 1903–1916, 1994.

Guzzi, R., Rizzi, R., and Zibordi, G., Atmospheric correction of data measured by a flying platform over the sea: elements of a model and its experimental verification, *Appl. Opt.*, 26, 3043–3051, 1987.

Häder, D.-P. and Worrest, R.C., Effects of enhanced solar ultraviolet radiation on aquatic ecosystems, *Photochem. Photobiol.*, 53, 717–725, 1991.

Häder, D.-P. (Ed.), *The Effects of Ozone Depletion on Aquatic Ecosystems,* Academic Press, San Diego, CA, 1997.

Haltrin, V.I. and Kattawar, G.W., Self consistent solutions to the equation of transfer with elastic and inelastic scattering in ocean optics. I. Model, *Appl. Opt.*, 32, 5356–5367, 1993.

Haras, W.S., Bukata, R.P., and Tsui, K.K., Methods for recording Great Lakes shoreline change, *Geosci. Can.*, 3, 174–184, 1976.

Harm, W., *Biological Effects of Ultraviolet Radiation,* Cambridge University Press, Cambridge, U.K., 1980.

Harris, G.P., The relationship between chlorophyll *a*, fluorescence, diffuse attenuation changes and photosynthesis in natural phytoplankton populations, *J. Plankton Res.*, 2, 109–127, 1980.

Hawes, S.K., Quantum fluorescence efficiencies of marine fulvic and humic acids, M.Sc. thesis, Dept. of Marine Sciences, U. of South Florida, St. Petersburg, FL., 92 pp., 1992.

Heath, D.F., Krueger, A.L., Roeder, H.A., and Henderson, B.D., The solar backscatter ultraviolet and total ozone mapping spectrometer (SBUV/TOMS) for Nimbus G., *Opt. Eng.*, 14, 323–326, 1975.

Hilton, J., Airborne remote sensing for freshwater and estuarine monitoring, *Water Res.*, 18, 1195–1223, 1984.

Højerslev, N.K., Spectral absorption by gelbstoff in coastal waters displaying highly different concentrations, in *Ocean Optics XIV, Proc. SPIE* (Ackleson, S.G. and Campbell, J. Eds.), Office of Naval Research, Washington, D.C., 1998.

Howard, K.L. and Yoder, J.A., Contribution of the subtropical ocean to global primary production, in *Space Remote Sensing of the Sub-Tropical Oceans* (Liu, C.-T., Ed.), Pergamon Press, New York, 157–168, 1997.

Hoyle, F., The great greenhouse controversy, *Energy Environ.*, 7, 349–356, 1996.

IOCCG (1998), *Minimum Requirements for an Operational Ocean-Colour Sensor for the Open Ocean* (Morel, A., Ed.), Reports of the International Ocean-Colour Co-ordinating Group, No. 1, IOCCG, Dartmouth, Canada, 46 pp.

IOCCG (1999), *Status and Plans for Satellite Ocean-Colour Missions: Considerations for Complementary Missions* (Yoder, J.A., Ed.), Reports of the International Ocean-Colour Coordinating Group, No. 2, IOCCG, Dartmouth, Canada, 43 pp.

IOCCG (2000), *Remote Sensing of Ocean Colour in Coastal, and Other Optically Complex Waters*, (Sathyendranath, S., Ed.), Reports of the International Ocean-Colour Coordinating Group, No. 3, IOCCG, Dartmouth, Canada, 140 pp.

IOCCG (2004), *Guide to the Creation and Use of Ocean Colour, Level 3, Binned Data Products* (Antoine, D., Ed.), Reports of the International Ocean-Colour Coor-dinating Group, No. 4, IOCCG, Dartmouth, Canada, 88 pp.

ISESS, *Environmental Software Systems, Environmental Information and Decision Support, International Federation for Information Processing TC5 EG5.11 Third International Symposium on Environmental Software Systems* (Denzer, R., Swayne, D.A., Pur-vis, M., and Schimak, G., Eds.), Kluwer Academic Publishers, Boston, MA, 268 pp., 1999.

Jaquet, J.-M., Schanz, F., Bossard, P., Hanselmann, K., and Beck, C., Measurements and significance of bio-optical parameters in two subalpine lakes of different trophic state in perspective of remote sensing, *Aquatic Sci.*, 56, 263–305, 1994.

Jerlov, N.G., Optical studies of ocean water, *Rep. Swedish Deep-Sea Exped.*, 3, 1–59, 1951.

Jerlov, N.G., Influence of suspended and dissolved matter on the transparency of sea water, *Tellus*, 5, 59–65, 1953.

Jerlov, N.G., *Marine Optics*, Elsevier Oceanography Series 14, Elsevier Publishing Co., Amsterdam, 231 pp., 1976.

Jerome, J.H., Bukata, R.P., and Bruton, J.E., Spectral attenuation and irradiance in Laurentian Great Lakes, *J. Great Lakes Res.*, 9, 60–68, 1983.

Jerome, J.H., Bukata, R.P., and Bruton, J.E., Utilizing the components of vector irra-diance to estimate the scalar irradiance in natural waters, *Appl. Opt.*, 27, 4012–4018, 1988.

Jerome, J.H., Bukata, R.P., and Miller, J.R., Remote sensing reflectance and its relationship to optical properties of natural waters, *Int. J. Remote Sens.*, 17, 3135–3155, 1996.

Jerome, J.H. and Bukata, R.P., Tracking the propagation of solar ultraviolet radiation: dispersal of ultraviolet photons in inland waters, *J. Great Lakes Res.*, 24, 666–680, 1998a.

Jerome, J.H. and Bukata, R.P., Impact of stratospheric ozone depletion on photoproduction of hydrogen peroxide in Lake Ontario, *J. Great Lakes Res.*, 24, 929–935, 1998b.

Keiner, L.E. and Yan, X.H., A neural network model for estimating sea surface chlorophyll and sediments from thematic mapper imagery, *Remote Sens. Environ.*, 66, 153–165, 1998.

Kininmonth, W., *Climate Change: A Natural Hazard*, Multi-Science Publishing Company, Ltd., Brentwood Essex, 310 pp., 2004.

Kirk, J.T.O., Spectral absorption properties of natural waters: contribution of the soluble and particulate fractions to light absorption in natural waters, *Aust. J. Mar. Freshwater Res.*, 27, 61–71, 1980.

Kirk, J.T.O., Monte Carlo procedure for simulating the penetration of light into natural waters, CSIRO Aust. Div. Plant Ind. tech. pap., 36, 1–16, 1981a.

Kirk, J.T.O., Monte Carlo study of the nature of the underwater light field in, and the relationship between optical properties of, turbid yellow waters, *Aust. J. Mar. Freshwater Res.*, 32, 517–532, 1981b.

Kirk, J.T.O., *Light and Photosynthesis in Aquatic Ecosystems*, Cambridge University Press, Melbourne, Australia, 401 pp., 1983.

Kirk, J.T.O., Dependence of relationship between inherent and apparent optical properties on solar altitude, *Limnol. Oceanogr.*, 29, 350–356, 1984.

Kirk, J.T.O., The upwelling light stream in natural waters, *Limnol. Oceanogr.*, 34, 1410–1425, 1989.

Kirk, J.T.O., *Light and Photosynthesis in Aquatic Ecosystems*, Cambridge University Press, U.K., 509 pp., 1994a.

Kirk, J.T.O., Optics of UV-B radiation in natural waters, *Archiv für Hydrobiologie*, 43, 1–16, 1994b.

Kishino, M., Sugihara, S., and Okami, N., Theoretical analysis of the *in situ* fluorescence of chlorophyll *a* on the underwater spectral irradiance, *Mer*, 21, 130–138, 1986.

Kneizys, F.X., Shettle, E.P., Gallery, W.O., Chetwynd, J.H., Abreu, L.W., Selby, J.E.A., Clough, A., and Fenn, R.W., *Atmospheric Transmittance/Radiance Computer Code LOWTRAN 6*, Air Force Geophysics Lab., Hanscom AFB Environmental Research Report AFGL-TR-83-0187, 1983.

Kopelevich, O.V. and Ershova, S.V., Model for seawater optical characteristics at UV spectral region, in *Ocean Optics XIII, Proc. SPIE 2963* (Ackleson, S.G. and Frouin, R. Eds.), Bellingham, WA, 167–172, 1997.

Kullenberg, G., Scattering of light by Sargasso Sea water, *Deep-Sea Res.*, 15, 423–432, 1968.

Lachance, M., Bobee, B., and Gouin, D., Characterization of the water quality in the St. Lawrence River: determination of homogeneous zones by correspondence analysis, *Water Resources Res.*, 15, 1451–1462, 1979.

Laegdsgaard, P., *A Field Guide for the Identification and Monitoring of the Seagrasses and Macroalgae in Wallis Lake*, Land and Water Conservation, Centre for Natural Water Resources, NSW Government, 2001.

Lee, Z.P., Carder, K.L., Steward, R.G., Peacock, T.G., Davis, C.O., and Mueller, J.L., Remote sensing reflectance and inherent optical properties of oceanic waters derived from above-water measurements, in *Ocean Optics XIII, Proc. SPIE,* 2963 (Ackleson, S.G. and Frouin, R., Eds.), 160–166, 1996a.

Lee, Z.P., Carder, K.L., Peacock, T.G., Davis, C.O., and Mueller, J.L., A method to derive ocean absorption coefficients from remote sensing reflectance, *Appl. Opt.,* 35, 453–462, 1996b.

Lee, Z.P., Carder, K.L., Steward, R.G., Peacock, T.G., Davis, C.O., and Patch, J.S., An empirical algorithm for light absorption by ocean water based on color, *J. Geophys. Res.,* 103, 27,967–27,978, 1998.

Lee, Z.P., Carder, K.L., Mobley, C.D., Steward, R.G., and Patch, J.F., Hyperspectral remote sensing for shallow waters. 2. Deriving bottom depths and water properties by optimization, *Appl. Opt.,* 38, 3831–3843, 1999.

Levenberg, K., A method for the solution of certain non-linear problems in least squares, *Quant. Appl. Math.,* 2, 164–168, 1944.

Lewis, M.R., Cullen, J.J., and Platt, T., Phytoplankton and thermal structure in the upper ocean: consequences of nonuniformity in chlorophyll profile, *J. Geophys. Res.,* 88, 2565–2570, 1983.

Light, B. and Beardall, J., Distribution and spatial variation of benthic microalgal biomass in Port Phillip Bay, Victoria, *Aquatic Botany,* 1010–1017, 1998.

Lillesand, T.M. and Kiefer, R.W., *Remote Sensing and Image Interpretation,* 4th ed., John Wiley & Sons, New York, 736 pp., 2000.

Lillesand, T.M., Riera, J., Chipman, J., Gage, J., Janson, M., Panuska, J., and Webster, K., Integrating multi-resolution satellite imagery into a satellite lake observatory, *Proc. Am. Soc. Photogramm. Remote Sens.,* St. Louis, MO, April, 2001.

Lindell, L.T., Pierson, D., Premazzi, G., and Zillioli, E., *Manual for Monitoring European Lakes Using Remote Sensing Techniques,* European Communities, Luxembourg, 164 pp., 1999.

Loisel, H. and Stramski, D., Estimation of the inherent optical properties of natural waters from the irradiance attenuation coefficient and reflectance in the presence of Raman scattering, *Appl. Opt.,* 39, 3001–3011, 2000.

Longhurst, A., Sathyendranath, S., Platt, T., and Caverhill, C., An estimate of global primary production in the ocean from satellite radiometer data, *J. Plankton Res.,* 17, 1245–1271, 1995.

Mann, M.E., Bradley, R.S., and Hughes, M.K., Global-scale temperature patterns and climate forcing over the past six centuries, *Nature,* 392, 779–787, 1998.

Mann, M.E., Bradley, R.S., and Hughes, M.K., Northern Hemisphere temperatures during the past millennium: inferences, uncertainties, and limitations, *Geophys. Res. Let.,* 26, 759–762, 1999.

Marquardt, D.W., An algorithm for least squares estimation of nonlinear parameters, *J. Int. Soc. Appl. Math.,* 11, 36–48, 1963.

Marshall, B.R. and Smith, R.C., Raman scattering and in-water ocean optical properties, *Appl. Opt.,* 29, 71–84, 1990.

Maul, G.A., *Introduction to Satellite Oceanography,* Martinus Nijhoff Publishers, Dordrecht, The Netherlands, 606 pp., 1985.

McCracken, K.G., Rao, U.R., and Bukata, R.P., Cosmic ray propagation processes. 1. A study of the cosmic ray flare effect, *J. Geophys. Res.,* 72, 4293–4324, 1967.

McIntyre, S. McKitrick, R., Hockey sticks, principal components, and spurious significance, to appear in *Geophys. Res. Let.,* 2005.

Michaels, P. and Knappenberger, P., Forging consensus: climate modeling and scientific review in the IPCC, in *The Global Warming Debate: The Report of the European Science and Environmental Forum*, London, 158–178, 1996.

Mobley, C.D., *Light and Water; Radiative Transfer in Natural Waters*, Academic Press, San Diego, CA, 592 pp., 1994.

Mobley, C.D. and Sundman, L.K., *HYDROLIGHT 4.1 Users' Guide*, Sequoia Scientific, Inc., Mercer Island, WA, 85 pp., 2000a.

Mobley, C.D. and Sundman, L.K., *HYDROLIGHT 4.1 Technical Documentation*, Sequoia Scientific, Inc., Mercer Island, WA, 76 pp., 2000b.

Morel, A., Inwater and remote measurements of ocean color, *Boundary-Layer Meteorol.*, 18, 177–201, 1980.

Morel, A., Optical modeling of upper ocean in relation to its biogenous matter content, *J. Geophys. Res.*, 93, 10749–10768, 1988.

Morel, A. and Prieur, L., Analysis of variations in ocean color, *Limnol. Oceanogr.*, 22, 709–722, 1977.

Morel, A. and Berthon, J.-F., Surface pigments, algal biomass profiles, and potential production of the euphotic layer: relationships reinvestigated in view of remote sensing applications, *Limnol. Oceanogr.*, 34, 1545–1562, 1989.

Morel, A. and Gentili, B., Diffuse reflectance of oceanic waters; its dependence on sun angle as influenced by the molecular scattering contribution, *Appl. Opt.*, 30, 4427–4438, 1991.

Morel, A. and Gentili, B., Diffuse reflectance of oceanic waters. II. Bidirectional aspects, *Appl. Opt.*, 32, 6864–6879, 1993.

Mourad, P.D., Footprints of atmospheric phenomena in synthetic aperture images of the ocean surface — a review, in *Air–Sea Exchange: Physics, Chemistry, and Dynamics* (Geernaert, G.L., Ed.), Kluwer Academic, Dordrecht, The Netherlands, 269–290, 1999.

Mueller, J.L., Ocean color measured off the Oregon coast: characteristic vectors, *Appl. Opt.*, 15, 394–402, 1976.

Müller–Karger, F.E., McClain, C.R., Fisher, T.R., Esaias, W.E., and Varela, R., Pigment distribution in the Caribbean Sea: observations from space, *Prog. Oceanogr.*, 23, 23–64, 1989.

Murray, J. and Hjort, J., *The Depth of the Ocean*, Macmillan, London, 821 pp., 1912.

Nelder, J.A. and Mead, R., A simplex method for function minimization, *Comput. J.*, 7, 308–313, 1965.

Neville, R.A. and Gower, J.F.R., Passive remote sensing via chlorophyll fluorescence, *J. Geophys. Res.*, 82, 3487–3493, 1977.

Okaichi, T., Anderson, D.M., and Nemoto, T. (Eds.), *Red Tides: Biology, Environmental Science and Toxicology*, Elsevier, New York, 1989.

O'Reilly, J.E., Maritorena, S., Mitchell, B.G., Siegel, D.A., Carder, K.L., Garver, S.A., Kahru, M., and McClain, C., Ocean color chlorophyll algorithms for SeaWiFS, *J. Geophys. Res.*, 103, 24937–24953, 1998.

Pan, Y., Subba Rao, D.V., and Warnock, R.E., Photosynthesis and growth of *Nitzschia pungens* f. multiseries Hasle, a neurotoxin producing diatom, *J. Exp. Marine Biol. Ecol.*, 154, 77–96, 1991.

Peacock, T.G., Carder, K.L., Davis, C.O., and Steward, R.G., Effects of fluorescence and water Raman scattering on models of remote sensing reflectance, in *Ocean Optics X, Proc. SPIE 1320*, 303–319, 1990.

Pelevin, V.N. and Rutkovskaya, V.A., On the optical classification of ocean waters from the spectral attenuation of solar radiation, *Oceanology*, 18, 278–282, 1977.

Pernetta, J.C. and Milliman, J.D., *Land–Ocean Interactions in the Coastal Zone Implementation Plan*, IGBP report No. 3, Stockholm, 215 pp., 1995.

Petzold, T.J., *Volume Scattering Functions for Selected Ocean Waters*, Scripps Institute of Oceanography Ref. 72-28, Univ. of CA, San Diego, 79 pp., 1972.

Phillips, D.M. and Kirk, J.T.O., Study of the spectral variation of absorption and scattering in some Australian coastal waters, *Aust. J. Mar. Freshwater Res.*, 35, 635–644, 1984.

Phillips, R.C. and McRoy, C.P. (Eds.), *Handbook of Seagrass Biology: an Ecosystem Perspective*, Garland STPM Press, New York, 353 pp., 1980.

Plass, G.N. and Kattawar, G.W., Monte Carlo calculations of radiative transfer in the Earth's atmosphere–ocean system: 1. Flux in the atmosphere and ocean, *J. Phys. Oceanogr.*, 2, 139–145, 1972.

Platt, T., Primary production in the ocean water column as a function of surface light intensity, *Deep-Sea Res.*, 33, 149–163, 1986.

Platt, T. and Irwin, B., Caloric content of phytoplankton, *Limnol. Oceanogr.*, 18, 306–310, 1973.

Platt, T. and Herman, A.W., Remote sensing of phytoplankton in the sea: surface chlorophyll as an estimate of water column chlorophyll and primary production, *Int. J. Remote Sens.*, 4, 343–351, 1983.

Platt, T. and Sathyendranath, S., Oceanic primary production: estimation by remote sensing at local and regional scales, *Science*, 241, 1613–1620, 1988.

Platt, T., Sathyendranath, S., Caverhill, C.M., and Lewis, M.R., Ocean primary production and available light: further algorithms for remote sensing, *Deep-Sea Res.*, 35, 855–879, 1988.

Platt, T., Harrison, W.G., Lewis, M.R., Li, W.K.W., Sathyendranath, S., Smith, R.E., and Vézina, A.F., Biological production of the oceans: the case for a consensus, *Mar. Ecol. Prog. Ser.*, 52, 77–88, 1989.

Platt, T., Caverhill, C., and Sathyendranath, S., Basin-scale estimates of oceanic primary production by remote sensing: the North Atlantic, *J. Geophys. Res.*, 96, 15147–15159, 1991.

Platt, T. and Sathyendranath, S., Biological production models as elements of coupled, atmosphere–ocean models for climate research, *J. Geophys. Res.*, 96, 2585–2592, 1991.

Platt, T., Fuentes–Yaco, C., and Frank, K.T., Spring algal bloom and larval fish survival, *Nature*, 23, 398–399, 2003.

Pope, R.M. and Fry, E.S., Absorption spectrum (380–700 nm) of pure water: II. Integrating cavity measurements, *Appl. Opt.*, 36, 8710–8723, 1997.

Pozdnyakov, D.V., Lyaskovsky, A.V., Tanis, F.J., and Lyzenga, D.R., Modeling of apparent hydrooptical properties and retrievals of water quality in the Great Lakes for SeaWiFS: a comparison with *in situ* measurements, *Proc. IGARRS '99 Conf., Hamburg, Germany*, 2, 1143–1147, 1999.

Preisendorfer, R.W., *Hydrologic Optics, Vol. I: Introduction*, U.S. Dept. of Commerce, Washington, D.C., 1976a.

Preisendorfer, R.W., *Hydrologic Optics, Vol. II: Foundations*, U.S. Dept. of Commerce, Washington, D.C., 1976b.

Preisendorfer, R.W., *Hydrologic Optics, Vol. III: Solutions*, U.S. Dept. of Commerce, Washington, D.C., 1976c.

Preisendorfer, R.W., *Hydrologic Optics, Vol. IV: Imbeddings*, U.S. Dept. of Commerce, Washington, D.C., 1976d.

Preisendorfer, R.W., *Hydrologic Optics, Vol. V: Properties*, U.S. Dept. of Commerce, Washington, D.C., 1976e.

Prieur, L. and Sathyendranath, S., An optical classification of coastal and oceanic waters based on the specific spectral absorption curves on phytoplankton pigments, dissolved organic matter, and other particulate materials, *Limnol. Oceanogr.*, 26, 671–689, 1981.

Putnam, E.S. (Ed.), *System Concept for Wide-Field-of-View Observations of Ocean Phenomena from Space,* Report of the SeaWiFS (Sea viewing wide-field-of-view sensor) Working Group, NASA Greenbelt, MD, 1987.

Rao, U.R., McCracken, K.G., and Bukata, R.P., Cosmic ray propagation processes. 2. The energetic storm particle event, *J. Geophys. Res.*, 72, 4325–4341, 1967.

Richardson, L.L., Bukata, R.P., Cullen, J.J., Thiemann, S., and Gitelson, A., Sensors and methods for remote sensing of nearshore and inland waters, *Backscatter,* winter issue, 26–31, 2002.

Roesler, C.S., Perry, M.J., and Carder, K.L., Modeling *in situ* phytoplankton absorption from total absorption spectra in productive inland marine waters, *Limnol. Oceanogr.*, 34, 1510–1523, 1989.

Ruddick, K., Ovidio, F., and Rijkeboer, M., Atmospheric correction of SeaWiFS imagery for turbid coastal and inland waters, *Appl. Opt.*, 39, 897–912, 2000.

Sanders, R. and Tabuchi, S., Decision support system for flood risk analysis for the River Thames, United Kingdom, *J. Am. Soc. PE&RS*, 66, 1195–1208, 2000.

Sathyendranath, S. and Morel, A., Light emerging from the sea — interpretation and uses in remote sensing, in *Remote Sensing Applications in Marine Science and Technology* (Cracknell, A.P., Ed.), D. Reidel Publishing Co., Dordrecht, The Netherlands, 323–357, 1983.

Sathyendranath, S. and Platt, T., Computation of aquatic primary production: extended formalism to include effect of angular and spectral distribution of light, *Limnol. Oceanogr.*, 34, 188–198, 1989a.

Sathyendranath, S. and Platt, T., Remote sensing of ocean chlorophyll: consequence of nonuniform pigment profile, *Appl. Opt.*, 28, 490–495, 1989b.

Sathyendranath, S., Hoge, F.E., Platt, T., and Swift, R.N., Detection of phytoplankton pigments from ocean colour: improved algorithms, *Appl. Opt.*, 33, 1081–1089, 1994.

Sathyendranath, S., Longhurst, A.R., Caverhill, C.M., and Platt, T., Regionally and seasonally differentiated primary production in the North Atlantic, *Deep-Sea Res.*, 42, 1773–1802, 1995.

Sathyendranath, S. and Platt, T., Analytic model of ocean color, *Appl. Opt.*, 36, 2620–2629, 1997.

Sathyendranath, S. and Platt, T., Ocean-colour model incorporating trans-spectral processes, *Appl. Opt.*, 37, 2216–2227, 1998.

Sathyendranath, S., Subba Rao, D.V., Chen, Z., Stuart, V., Platt, T., Bugden, G.L., Jones, W., and Vass, P., Aircraft remote sensing of toxic phytoplankton blooms: a case study from Cardigan River, Prince Edward Island, *Can. J. Remote Sens.*, 23, 15–23, 1997.

Secchi, A., Relazione della esperienze fatta a bordo della Pirocorvetta L'Immacolata Concezione per determinaire la transparenza del mare. In Cialdi, A. *Sul moto ondoso del mare e su le correnti di esso specialment auquelle littorali,* 2nd ed., 255–288, 1866. (Translation available, Dept. of the Navy, Office of Chief of Naval Operations, O.N.I. Trans. No. A-655, Op-923M4B, Dec., 1955).

Setlow, R.B., The wavelengths in sunlight effective in producing skin cancer, *Proc. Nat. Acad. Sci. USA.*, 71, 3363–3366, 1974.

Shearer, J.A., DeBruyn, E.R., DeClercq, D.R., Schindler, D.W., and Fee, E.J., *Manual of Phytoplankton Primary Production Methodology*, Can. Tech. Rep. Fish. Aquat. Sci. No. 1341, 58 pp., 1985.

Sikora, T.D., Young, G.S., Beal, R.C., and Edson, J.B., Use of ERS-1 synthetic aperture radar imagery of the sea surface in detecting the presence and structure of the convective marine atmospheric boundary layer, *Mon. Wea. Rev.*, 123, 3623–3632, 1995.

Simonovic, S.P., Decision support system for flood management in the Red River basin, *Can. Water Res. J.*, 24, 203–233, 1999.

Singer, S.F., *The Scientific Case against the Global Climate Treaty*, Report of the Science & Environmental Policy Project (SEPP), 38 pp., 1997.

Singer, S.F., *Hot Talk, Cold Science: Global Warming's Unfinished Debate*, The Independent Institute, Oakland, CA, 120 pp., 1999.

Smayda, T. and Shimizu, Y. (Eds.), *Toxic Phytoplankton Blooms in the Sea*, Elsevier, Amsterdam, 1993.

Smith, E.L., Photosynthesis in relation to light and carbon dioxide, *Proc. Natl. Acad. Sci., USA*, 22, 504–511, 1936.

Smith, R.C. and Baker, K.S., Optical classification of natural waters, *Limnol. Oceanogr.*, 23, 260–267, 1978.

Smith, R.C. and Baker, K.S., Optical properties of the clearest natural waters (200–800 nm), *Appl. Opt.*, 29, 177–184, 1981.

Smith, R.C. and Baker, K.S., Stratospheric ozone, middle ultraviolet radiation and phytoplankton productivity, *Oceanogr. Mag.*, 2, 4–10, 1989.

Smith, R.C. and Wilson, W.H., Ship and satellite bio-optical research in the California Bight, in *Oceanography from Space* (Gower, J.F.R., Ed.), Plenum Press, New York, 1981.

Stainton, M.P., Capel, M.J., and Armstrong, F.A.J., *The Chemical Analysis of Fresh Water*, 2nd ed., Can. Fish. Mar. Serv. Misc. Spec. Publication No. 25, 180 pp., 1977.

Stavn, R.H. and Weidemann, A.D., Shape factors, two-flow models, and the problem of irradiance inversion in estimating optical parameters, *Limnol. Oceanogr.*, 34, 1426–1441, 1989.

Steffen, M., van Dam, W., Hogg, T., Breyta, G., and Chuang, I., Experimental implementation of an adiabatic quantum optimization algorithm, *Phys. Rev. Lett.*, 90, 0679031–0679034, 2003.

Stramski, D., Wozniak, S.B., and Flatau, P.J., Optical properties of Asian mineral dust suspended in seawater, *Limnol. Oceanogr.*, 49, 749–755, 2004.

Strickland, J.D.H., Production of organic matter in primary stages of the marine food chain, in *Chemical Oceanography*, (Riley, J.P. and Skirrow, G., Eds.), Academic Press, London, 477–610, 1965.

Strickland, J.D.H. and Parsons, T.R., *A Practical Handbook of Seawater Analysis*, Bull. Fish. Res. Board Can., 167, 310 pp., 1972.

Sugihara, S., Kishino, M., and Okami, N., Contribution of Raman scattering to upward radiance in the sea, *J. Oceanogr. Soc. Jpn*, 40, 397–404, 1984.

Sugumaran, R., Davis, C., Meyer, J., Prato, T., and Fulcher, C., Web-based decision support tool for floodplain management using high-resolution DEM, *J. Am. Soc. PE&RS*, 66, 1261–1265, 2000.

Talling, J.F., The phytoplankton population as a compound photosynthetic system, *New Phytol.*, 56, 287–295, 1957.

Townsend, P.A., and Walsh, S.J., Modeling floodplain inundation using an integrated GIS with radar and optical remote sensing, *Geomorphology,* 21, 295–312, 1998.

Tyler, J.E., The Secchi disc. *Limnol. Oceanogr.,* 13, 1–6, 1968.

Unoki, S., Okami, N., Kishino, M., and Sugihara, S., *Optical Characteristics of Sea Water at Tokyo Bay,* Science and Technology Agency Report, Japan, 1978.

Villafane, V.E., Helbling, E.W., Holm–Hansen, O., and Chalker, B.E., Acclimatization of Antarctic natural plankton assemblages when exposed to solar ultraviolet radiation, *J. Plankton Res.,* 17, 2295–2306, 1995.

Vodacek, A., Green, S.A., and Blough, N.V., An experimental model of the solar-stimulated fluorescence of chromophoric dissolved organic matter, *Limnol. Oceanogr.,* 39, 1–11, 1994.

Vollenweider, R.A., Some observations on the C14 method for measuring primary production, *Verh. Internat. Verein. Limnol.,* 14, 134–139, 1961.

Vollenweider, R.A., Calculation models of photosynthesis-depth curves and some implications regarding day rate estimates in primary production measurements, *Mern. Inst. Ital. Idrobiol.,* 18 (suppl), 425–457, 1965.

West, R.J., Thorogood, C., Walford, T., and Williams, R.J., *An Estuarine Inventory for New South Wales, Australia,* Department of Agriculture, NSW, 1985.

Whitehouse, B.G., Defining "too expensive," editorial, *Backscatter,* winter issue, 4, 2002.

Whiting, J., Determination of groundwater inflow to prairie lakes using remotely sensed data, *Proc. Symp. Machine Processing Remotely Sensed Data, IEEE Trans. Geosci. Electronics,* GE-14, 1, 60–65, 1976.

Whiting, J., The effect of groundwater on evaporation from a saline lake, *J. Appl. Meteorol.,* 23, 214–221, 1984.

Whiting, J. and Bukata, R.P., An examination of the role of satellites in monitoring the impact of climate change on the hydrological cycle, *Proc. Int. Soc. Photogrammetry Remote Sensing Commission, VII Symp. Global Environ. Monitoring,* 2, Sept., Victoria, B.C., 10 pp., 1990.

Whiting, J. and Bukata, R.P., Monitoring the global change of water resources using remote sensing, in *The Canadian Remote Sensing Contribution to Understanding Global Change* (LeDrew, E., Strome, M., and Hegyi, F., Eds.), University of Waterloo, Publication Series 38. chapter 10, 199–236, 1995.

Whitlock, C.H., Poole, L.R., Ursy, J.W., Houghton, W.M., Witte, W.G., Morris, W.D., and Gurganus, E.A., Comparison of reflectance with backscatter for turbid waters, *Appl. Opt.,* 20, 517–522, 1981.

Williamson, C.E. and Zangarese, H.E. (Eds.), *Advances in Limnology: Impact of UV-B Radiation on Pelagic Freshwater Systems,* Archiv für Hydrobiologie Special Issue 43, 226 pp., 1994.

Worrest, R.C., Impact of solar ultraviolet-B radiation and aquatic primary production: damage, protection and recovery, *Physiol. Plant.,* 58, 428–434, 1983.

Wrigley, R.C., Spanner, M.A., Slye, R.E., Pueschel, R.F., and Aggarwal, H.R., Atmospheric correction of remotely sensed image data by a simplified model, *J. Geophys. Res.,* Special FIFE issue, 97, 18,797–18,814, 1992.

Yoo, S-J. and Jeong, J-C., Detecting red tides in turbid waters, *J. Korean Soc. Remote Sens.,* 15, 321–327, 1999.

Index

A

Printed and bound by CPI Group (UK) Ltd, Croydon, CR0 4YY

28/10/2024

01780249-0001